A GUIDE TO Carnivorous Plants of the WORLD

GORDON CHEERS

A GUIDE TO CARNIVOROUS PLANTS of the WORLD

GORDON CHEERS

Angus&Robertson
An imprint of HarperCollins*Publishers*

For Julie, with love

AN ANGUS & ROBERTSON BOOK
An imprint of HarperCollinsPublishers

First published in Australia in 1992 by
CollinsAngus&Robertson Publishers Pty Limited (ACN 009 913 517)
A division of HarperCollinsPublishers (Australia) Pty Limited
25-31 Ryde Road, Pymble NSW 2073, Australia

HarperCollinsPublishers (New Zealand) Limited
31 View Road, Glenfield, Auckland 10, New Zealand

HarperCollinsPublishers Limited
77-85 Fulham Palace Road, London W6 8JB, United Kingdom

Distributed in the United States of America by
HarperCollinsPublishers
10 East 53rd Street, New York NY 1002

Copyright © Gordon Cheers 1992
Copyright © Margaret Hodgson, illustrations 1992
Copyright for photographs remains with individual photographers, as per picture credits listed in back of book.

This book is copyright.

Apart from any fair dealing for the purposes of private study, research, criticism or review, as permitted under the Copyright Act, no part may be reproduced by any process without written permission. Inquiries should be addressed to the publishers.

National Library of Australia
Cataloguing-in-Publication data:

Cheers, Gordon
  A guide to carnivorous plants of the world.
  Bibliography.
  Includes index.
  ISBN 0 207 16186 0.
  1. Carnivorous plants I. Title.
583.121

Cover photograph by Andre Martin
Special thanks to the Royal Botanical Gardens Sydney
for supplying the plant for the cover photograph.

Typeset in Garamond by Midlands Typesetters, Australia
Printed in the People's Republic of China.
5 4 3 2 1
96 95 94 93 92

Contents

Acknowledgments vi

Preface vii

Climate Key ix

Pronunciation Guide ix

Introduction 1

History, Fact and Folklore 5

Evolutionary Paths 16

Parts of a Carnivorous Plant 18

Classification of Carnivorous Plants 21

Trapping Mechanisms 23

Cultivation 29

Propagation 39

Carnivorous Plant Genera 48

Field Trips 131
   Albany, Western Australia 131
   Anglesea, Victoria 133
   Arthur's Pass National Park, New Zealand 135
   Mount Kinabalu, Borneo 139
   Mount Roraima, Venezuela 142
   North America 144

Glossary 145

Monthly Calendar 146

Cultivating Guidelines 150

World Carnivorous Plant List 152

References 167

Recommended Reading 167

Organisations 168

Bibliography 169

Picture Credits 172

Index of Scientific Names 173

Acknowledgments

I would particularly like to thank my ever-patient friend Larry Pitt, who took many photographs for this book; Stefanie Hamel and Helen Gregory, who struggled with my bad handwriting; Patricia Pollard who helped capture the characteristics of plants with her pen, and Margaret Hodgson who finally produced the illustrations; my patient and enthusiastic editor Kate Evans; Alan Pollard, who was there at the beginning, and Julie Silk, who forced me to finish.

Over many years commercial growers, carnivorous plant enthusiasts, botanists and others have provided contributions to make this book possible. Some have provided hints on how to grow better plants, some have brought recent articles to my attention, and others still have provided information on the long-term development of plants in their natural habitat. These people include the following:

Australia: David Bond; Jenny Brownfield; Paul Cain; Steve Clemesha; Richard Davion; Kingsley Dixon; Brian Denton; Ian English; John Forlonge; Phillip Harris; Fred Howell; Steve Jackson; Peter Lavarack; Alan Lowrie; Phil Mann; Bruce Pierson; Geoff Roberts; Laurie Ritter; Ian Rogers; Roger Shivas; Justin Tong; Peter Tsang; Adrian Walter; Brian Whitehead

Canada: Ann Middleton; John Turnbull

Czechoslovakia: Miloslav Studnicka

France: Alain Christophe; Marcel Lecoufle

Germany: Thomas Alt; Josef Bogner; Hugo Herkner; Johannes Marabini

Japan: Hiroshi Fukatsu; Yasuhiro Fukatsu; A. Hayashi; Katsuhiko Kondo; Isamu Kusakabe

Malaysia: Anthony Lamb; Robert Cantley

New Zealand: Philip Cotter

Switzerland: Jacques Haldi; Rudo Schmid-Hollinger; Lorenz Bütschi

USA: Richard M. Adams II; Gordon Blanz; Donald Clements; Cliff Dodd; Patricia Dwyer; Joe Mazrimas; Donald Schnell; Richard Silversten; Steve Smith; Warren Stoutamire; Steven Williams

UK: J.K. Burras; Eric Binstead; Barry Juniper; Adrian Slack; David Taylor; Peter Taylor; John Watkins

I would like to reiterate my particular thanks to photographer Larry Pitt, who has so beautifully captured the vitality of carnivorous plants. His photographic skill, and patience in travelling to remote locations and working under difficult circumstances, are much appreciated. Larry happily spent time setting up complicated shots and photographing plants at various stages over a number of months, demonstrating visually what could never be described in words. My thanks.

Preface

When I first started growing carnivorous plants, I wondered what made these plants different to 'normal' plants. Why were these 550 or so species of plants carnivorous when most of the plant world is quite happy to have a meat-free diet? Now, many years on, after growing and studying these fascinating plants, I no longer wonder why these plants are carnivorous—I wonder why the rest of the plant world is *not* carnivorous.

Having earnt my living by growing and selling carnivorous plants, I have had the opportunity to communicate with hobbyist and commercial growers from all over the world. Together, we have discussed the many different methods of cultivating and propagating carnivorous plants. I have described these methods in this book, and the diversity of methods reflects the diversity of carnivorous plants and their habitats. Because of the wide range of habitats—there are carnivorous plants in almost every area of the world—I have described not only individual species in the genera section, but also the environment in which they are found. By understanding the similarities and differences in environment and surrounding vegetation for different species, the carnivorous plant grower not only gains a better understanding of the plants themselves, but is also more able to duplicate, and even improve upon, the conditions in which they grow.

I have observed many carnivorous plants in their native habitat, and have tried to create similar environments for them when growing them myself. I have had success with even the most rare and difficult of species: I hope this book helps and inspires you to do the same.

Utricularia cornuta

Drosera oreopodion

Climate Key

When describing climate, I have used four temperature ranges, as follows. These ranges are general, and more specific temperature ranges are indicated within genus and species descriptions, where relevant. I have included a map indicating temperature ranges as they relate to geographical locations, but of course this also changes according to altitude and other factors. Again, specific temperature ranges will be indicated where necessary.

Tropical   Average temperature 20–30°C (68–86°F)
Subtropical Average temperature 15–20°C (59–68°F)
Temperate   Average temperature 10–15°C (50–59°F)
Boreal   Average temperature 0–10°C (32–50°F)

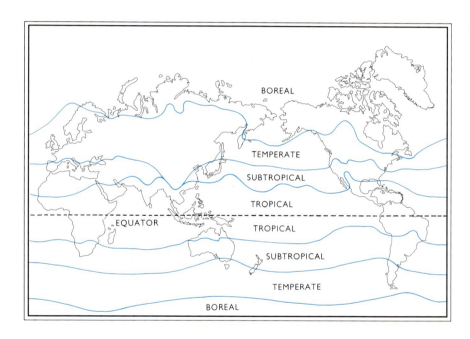

Pronunciation Guide

A phonetic guide to the pronunciation of scientific names has been given for every genus within the Carnivorous Plant Genera section. This follows simple principles regarding length of vowels and emphasis on syllables.

- ˙ short vowel
- ˘ medium or 'regular' length vowel
- ¯ long vowell
- ˈ major stress
- ˌ secondary stress

Sarracenia x catesbaei

Introduction

Utricularia purpurea

Carnivorous plants share one characteristic, that is the ability to capture prey, dissolve and gain nourishment from it. Apart from this, they must be the most diverse group of plants in existence.

The carnivorous plants group consists of nine families, 17 genera and over 550 species. There are species native to almost every area of the world, from snow-covered alpine regions to tropical jungles. Even within some genera, there is massive diversity of natural habitat. *Drosera*, for example, has species such as *D. arcturi*, found on the New Zealand alps; *D. peltata*, which in its native habitat grows on ground that is hard, dusty and dry during the hot summer; and *D. prolifera*, which is native to tropical Australia. The *Nepenthes* genus, with 70 or so species, has pitcher shapes that vary from the large football-like pitcher of *N. rajah*, reputedly capable of capturing rats, frogs and birds, to the tiny pitchers of *N. tentaculata* that capture nothing larger than an ant. In contrast, there are monotypic genus whose single species are endemic only to a small and ever-decreasing region, as is the case with *Darlingtonia californica* and *Cephalotus follicularis*: apart from being endangered and carnivorous, these two species have little in common. *Aldrovanda vesiculosa*, another monotypic species, has similar trapping mechanisms and floral parts to *Dionaea*, with a notable distinction—*A. vesiculosa* is aquatic, with its traps below the water line and its small white flower floating above the surface while *Dionaea* is terrestrial.

Differences between carnivorous plants are more complex than merely cosmetic—actual mechanisms and methods of being carnivorous vary from genus to genus. Trapping methods show an interesting variety. The well-known *Dionaea muscipula* (Venus Fly Trap) has its distinctive traps at the end of its leaves, while *Genlisea* has a trap consisting of two spiral twisting arms with multiple entrances that lead the prey on into the digestive centre; *Utricularia* has a trapping system that involves mucilage, sugar-producing cells, trigger hairs and a vacuum system, and members of the *Byblis* genus have sticky leaves that hold the prey.

Some genera do, however, share characteristics. *Heliamphora*, found on remote plateaus in tropical Venezuela, have a similarly shaped pitcher to *Sarracenia*. They both have downward-pointing hairs that guide prey inward and to their death, and they are also the only two to produce a real scent. Their differences lie in the size of pitchers: *Heliamphora* pitchers are generally shorter and fatter than those of *Sarracenia*, and they have a protective lid that is quite small, and has almost disappeared, as if it was an outmoded appendage. While the noddling flowers of *Darlingtonia*, *Sarracenia* and *Heliamphora* are similar, and grow to no more than 18 cm (7.1 in) in height, those on other carnivorous plants, such as the tropical pitcher plant *Nepenthes*, have a totally different inflorescence, which can grow up to 1 m (3 ft) high.

Carnivorous plants are clearly diverse, and they are also,

in a sense, mysterious and exotic. They are also thought to be difficult to grow, but for most species this is not the case, as is indicated throughout this book. To consolidate this information, I have provided a cultivation table with simple guidelines for both the novice and the experienced grower. It covers areas such as types of soil to use, watering methods, light requirements and degree of difficulty. There is also a monthly calendar that details yearly growing patterns of major species. These two tables, together with Cultivation, Propagation and Trapping chapters, will make it possible to understand a particular genus more fully, and to adjust growing conditions accordingly, to produce better plants.

I have also included a section describing field trips all over the world, searching for carnivorous plants. Understanding natural environments and surrounding vegetation is helpful for both the novice and the professional, as trying to emulate a plant's natural environment is the surest path to success with your carnivorous plants.

At the end of this book I have included a comprehensive listing of carnivorous plants of the world, showing all genera, species and hybrids, and known synonyms. The carnivorous plant family is a changing one, and genus are often reassessed, *Triphyophyllum*, *Brocchinia*, *Catopsis* and *Ibicella* being good examples of this. Further, with more botanists understanding in greater detail the way carnivorous plants capture and absorb their prey, and with more enthusiasts developing a deeper understanding of these plants, it is likely that more plants will be recognised and classified as carnivorous, and the World Plant List will continue to change and develop.

DIVERSITY AND *DROSERA*

Carnivorous plants are diverse and interesting: to illustrate this, here are a series of *Drosera*, showing scrambling, climbing, rosette, fan-leafed, erect and pygmy forms, all within the one genus.

Scrambling Drosera

Climbing Drosera

INTRODUCTION

Fan-leafed Drosera

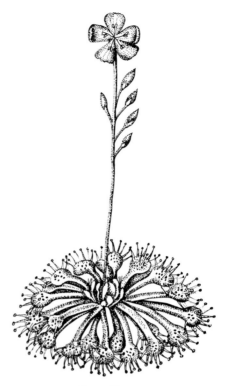

Pygmy Drosera

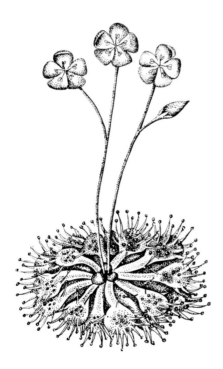

Rosette Drosera

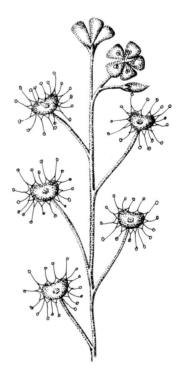

Erect Drosera

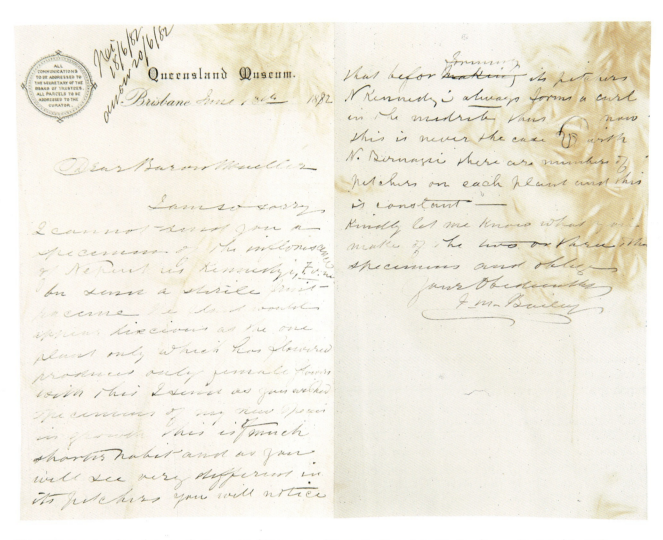

This 1882 correspondence between the Queensland Museum and Baron Ferdinand von Mueller, former director of the Melbourne Botanic Gardens and Government Botanist, indicates connections within the scientific community, and historic interest in carnivorous plants. In this letter, they are discussing the classification of the dried Nepenthes *species that appears on page 15.*

History, Fact and Folklore

Plants that think, hear, smell, move and eat? Mobile monsters—pursuing, trapping, torturing their prey, and eventually consuming their hapless victims: the thought of plants that eat human beings has captured our imagination for centuries. Many stories have been passed down from generation to generation, books written, and more recently spectacular movies made of fragile humankind falling into the mouth of carnivorous plants. Amid screams of pain and horror, the individual is dissolved away and the plant grows noticeably larger from the nourishment it has gained.

The reality is somewhat different. There are many plants that do gain a benefit from killing and absorbing living animals—these plants are carnivorous but not voracious monsters. Instead they are plants that tend to move quite slowly, almost imperceptibly; they are normal plants that you might have in your own home. Many 'normal' plants have one or more features common to carnivorous plants. Common species of vegetables are already known to trap insects and could, with slight variation, be carnivorous.

What, then, distinguishes a carnivorous plant from other plants is the ability to trap and gain nourishment from insects and small animals? The plant must first trap the prey, which it can do in a variety of ways. Next, it uses digestive enzymes to dissolve the victim, and cells within the plant absorb the resultant nourishing liquid in a process similar to human digestion. These processes usually happen as a result of the plant being alerted to the presence of its prey, to which it responds in order to gain nourishment.

The Tree That Ate Madagascar

Myths and legends about people-eating trees were plentiful in Europe in the late nineteenth and early twentieth centuries, especially about plants originating in the newly explored mysterious and exotic tropical lands such as Madagascar, New Guinea and Borneo. These legends came to prominence in 1860 when Carle Liche wrote an account of a female sacrifice fed to a tree. He described the tree as looking something like the top of a large pineapple, over 2 m (8 ft) high. This report was published in several European scientific publications and then in the New York *World*. It is not surprising that Liche's account received such

This imagined and extravagent 'carnivorous plant of the future' first appeared in an article by Leonard Bastin in Scientific America, *1909.*

worldwide publicity, given his graphic description. Liche describes how the natives fed a female sacrifice to the plant and began 'a grotesque and indescribably hideous orgy', and that the next day he found 'nothing but a white skull at the foot of the tree to remind me of the sacrifice that had taken place'.

Liche's narrative was not backed up by plant specimens, as is the standard procedure with all new plant discoveries. No real proof ever existed of the Madagascar Tree. Such tales must have really entertained the imagination of the Victorians, particularly as this was a period when distant tropical islands were being described to Europeans for the first time.

TRIFFIDS

More recently, in 1951, John Wyndham wrote *The Day of the Triffids*, giving us plants that stand up to 3 m (10 ft) tall and attack people with the intention of wiping out the human race in order to free the earth for Triffids alone. Much of the Triffids' features are based on real carnivorous plants.

Triffids constructed for the BBC television version of John Wyndham's The Day of the Triffids.

In Wyndham's book, these totally fictitious garden plants had, in their immature stage, a sticky centre similar to *Drosera* that held and digested insects. As the plants grew older they hobbled along the ground in a clumsy but effective manner. The shape of the Triffids was conceived from carnivorous plant species *Sarracenia flava*, with a few notable variations, the most spectacular of which is that the Triffids had a tongue-like appendage that sprang out at unsuspecting victims, stinging them with deadly poison.

In *The Day of the Triffids* the narrator and a small population of survivors were left in a world struck by disaster and almost totally overrun by these creatures, whose presence was preceded by an ominous, continuous clicking sound—the Triffids communicating with one another. This sound foreshadowed the 'eating' of a decomposed victim. Possibly reflecting this, and indicating the ways in which scientific reality and fictionalised knowledge influence each other, from within the *Sarracenia* you can hear the frantic buzzing of an insect as it throws itself against the sides of the pitcher, in a vain attempt to escape. When the sound inside the *Sarracenia* ceases, decomposition is not far away.

THE VOODOO LILY

Whether a plant is carnivorous or not has always been a debatable point. Some plants considered carnivorous in the past have since been proven non-carnivorous, for instance the Voodoo Lily. The Voodoo Lily (*Sauromatum guttatum*), a member of the Araceae family, uses insects to pollinate its flowers in an almost-carnivorous manner. The process is complicated, but to put it simply, the plant has a flower spike (or spadix) with male flower parts at the top of the plant and the female parts at the base, with a neutral zone in between. The female flowers mature before the male, making self-pollination impossible. The plant has overcome this pollination problem by giving off a pungent odour that attracts insects. Tiny droplets of oil coating the inside of the flower cause insects to fall to the base of the flower, and small bristles prevent them from leaving. At the base of the lily the insects become covered with a sweet-tasting liquid, then during the night the liquid is coated with falling pollen from the male flower. The barrier of bristles begins to wither making a fast exit now possible. The insect escapes, only to be attracted to the base of another lily, touching its female flower and fertilising a new flower.

Hence, although insects may be found buzzing around the base of a flower, and can be lured and trapped inside the plant and covered with a sweet-tasting liquid, the prison is only temporary—the plant has no ability to absorb its prey, and is not carnivorous. The same may be said for other plants, such as the Cruel Plant (*Physianthus albens*). Native to Brazil, it produces a flower in spring that attracts insects which, upon tasting the nectar, are lured along an ever-narrowing path and finally led to an adhesive to which the tongue of the insect sticks, and is held for life. This plant, like the Voodoo Lily, does not have the ability to dissolve its prey.

HISTORY, FACT AND FOLKLORE

Amorphophallus riviera, *family* Araceae, *was also initially and wrongly classified as carnivorous.*

Gronovia scandens, like the Cruel Plant, also holds insects, only this time the plant has barbed bristles so strong they can hold a reptile, that gradually dies from starvation. It cannot dissolve and digest its prey, as it has not evolved enough characteristics to be carnivorous.

ATTACK OF THE HOUSE PLANTS?

Writers such as Leonard Bastin have speculated on whether house plants will some day attack humans and take over the world, worrying that more and more plants are becoming carnivorous and that something must be done to prevent it. In 1909, in *Scientific American*, Bastin suggested that the carnivorous plants of today could be the 'man-eaters' of the future. He speculated that the carnivorous habit could not only become more widespread, but that our present carnivorous plants could also become much larger. The article depicts a giant sundew catching large birds such as storks, butterworts that could hold and later dissolve goats and other herd animals, large underwater bladderworts capable of devouring crocodiles, and so on. He considers how impossible it would be for anyone to escape from a large *Cephalotus follicularis*, given its slippery sides and pointed teeth. The article concludes by stating:

all natural changes come about with great slowness . . . the condition of man himself will have changed . . . otherwise the outlook for the human race is distinctly disquieting.

Recently, the rather remarkable discovery of a carnivorous genus, *Triphyophyllum*, has bought Bastin's predictions closer to reality. It is possibly a link between *Drosera*, *Nepenthes* and the common passionfruit. The species *Triphyophyllum peltatum* has been known for many years. As far back as 1951 suggestions were made that the plant closely resembled *Nepenthes*, yet also had gland-bearing leaves similar to *Drosera*, therefore belonging to one or other of those families. Nowhere was it suggested that the plant might be carnivorous until, in 1979, Green and Heslop-Harrison published a paper in which they described *Triphyophyllum* as 'a new carnivorous plant genus'. It goes through a series of changes, from a non-carnivorous plant

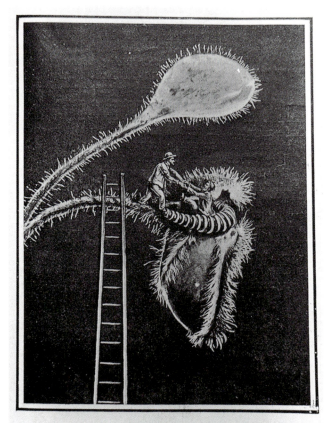

From a giant pitcher plant a man could escape only with the help of a friend.

Another of Leonard Bastin's theoretical house plants of the future, discussed in his Scientific America *article in 1909. Carnivorous plants have long excited speculation, imagination, and humour.*

to a carnivorous one, then back to a 'normal' state. Interestingly, it is carnivorous prior to flowering and also when damaged, when it requires most of its nourishment. Could this plant be the link between carnivorous and non-carnivorous plants that botanists such as F.E. Lloyd were looking for? I cannot say, as evidence is still incomplete.

Perhaps *T. peltatum* will eventually become solely carnivorous . . . if this is possible, what of the numerous plants today that capture and hold insects? What of our present vegetable crops, already able to ensnare pests?

Studies in the late 1970s have proved that on the foliage of both potato (*Solanum polyadenium*) and tomato plants (*Lycopersicon esculentum*), insects have been caught. The insects were trapped and died from starvation within eight days. Neither potatoes nor tomatoes have been known to digest the insects.

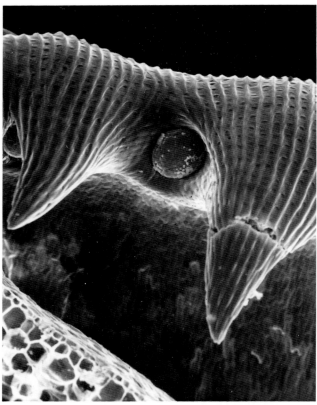

Electron microscopes have allowed even closer and more intense viewings of carnivorous plants. This photograph shows the pitcher orifice of a Nepenthes rafflesiana. *The cone-shaped structures are peristome 'teeth', separated by nectar glands. While reaching this nectar, insects fall from the slippery peristomes into the pitcher.*

Potato leaves have glandular hairs, containing a droplet of quick-setting liquid which, when ruptured, can hold an insect (for example an aphid) to the plant. These hairs could act as self-regulating 'insecticides', trapping disease-carrying insects.

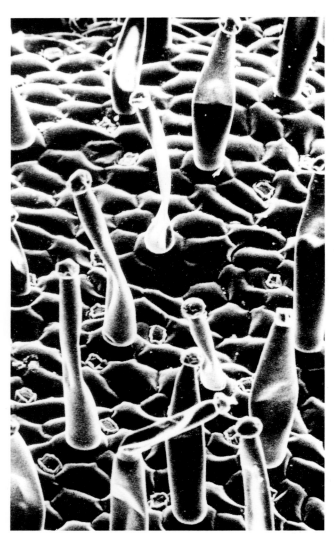

Electron microscope shot of the upper leaf surface of Pinguicula moranensis, *showing stalked mucilage-secreting glands, sessile glands that secrete digestive enzymes, and epidermal cells. The mucilage secreted traps small insects, while the protein-digesting enzymes from the sessile glands digest the prey.*

With the tomatoes that were engineered it was found that some leaves had from two to ten insects trapped on the upperside and the underside of the leaves. As with potato plants the glandular hairs on tomatoes may reduce the spread of leaf curl virus. Further work in this area is being planned. It may be possible to improve some varieties genetically so that more trapping hairs exist, and those that currently exist may be improved to catch a greater variety of insects. Where this will lead us is difficult to determine but possibly, before too long, all our vegetable crops will have some built-in mechanism to protect them from insects. Given current developments in modern plant hybridisation (protoplast fusion, cloning etc), large-scale vegetable crops that gain nourishment from trapped insects—carnivorous vegetables—may not be too far into the future.

GRANDFATHER AND GRANDSON DARWIN AND OTHERS

While some individuals have concerned themselves with hypothetical carnivorous plants for the future, or imaginary carnivorous plants from the past, others—such as Charles Darwin and J.D. Hooker—studied actual plants. They looked at species that were more than just speculative, and that have proven to be carnivorous.

Dr Erasmus Darwin, grandfather of Charles Darwin, suggested in the mid-eighteenth century that plants may be able to respond, of their own free will, to certain stimuli. So impressed was he with plants, particularly insectivorous plants, that he wrote the poem *The Botanical Garden*, which was an immediate success. It was published in 1791, when botany was becoming an accepted and even 'vogue' subject. It is a delightful poem, with some quite explicit erotic imagery that presumably both shocked and titillated the reader.

> *Queen of the marsh, imperial* Drosera *treads*
> *Rush-fringed banks, and moss-embroider'd beds;*
> *Redundant folds of glossy silk surrounds*
> *Her slender waist, and trail upon the ground;*
> *Five sister-nymphs collect with graceful ease,*
> *Or spread the floating purple to the breeze;*
> *And five fair youths with duteous love comply*
> *With each soft mandate of her moving eye.*

The poem expresses, in an imaginative and cunning way, his love of plants and his belief that there was much more to them than meets the human eye.

Charles Darwin

Ninety years later Charles Darwin, grandson of Erasmus, wrote *The Origin of Species* and, later, *Insectivorous Plants*. Many contemporaries of Darwin criticised his work, including *Insectivorous Plants*, and it is quite clear today that it was 'ahead of its time'. In 1881 one critic said, in *Scientific America*, that the plants were not carnivorous in any sense of the word and the theories should be taken *cum grano salis* (with a grain of salt). During the following year, however, evidence from other botanists began to prove Darwin's critics wrong and gradually confirm what had been written. Today *Insectivorous Plants* still stands as one of the great work on carnivorous plants, particularly on *Drosera*.

Like his grandfather, Charles was so taken with *Drosera* that whenever he had any leisure time he would spend it on experiments with the plants. Sixteen years after his first observation he wrote *Insectivorous Plants*, at which stage in his life he said of his favourite plant, 'I care more about the *Drosera* than the origin of all the species in the world'. We are left with the impression that Darwin considered the plant almost an animal with a nervous system. In a letter to his friend Asa Gray he wrote:

> *I can paralyse one half of the leaf so that a stimulus to the other causes no movement . . . just like dividing the spinal marrow of a frog . . . no stimulus can be sent from the brain.*

Darwin corresponded with botanists from all over the world, seeking information and plant specimens. This was a long and time-consuming correspondence. It could take two months or more before a reply was received from other lands. More immediate help came from his long-time friend J.D. Hooker who, living in England and working at Kew Gardens, was of invaluable assistance.

Joseph Hooker attained directorship of Kew Gardens in 1865 and travelled widely overseas, bringing back with him both plants and information, making Kew Gardens the centre of exchange for botanic information not only for the British Empire, but for the whole world. Even today Kew Gardens makes headlines with its micropropagation ('test-tube plants') and sends expeditions to areas where botanical information about the flora is still largely unpublished, such as the summits of South East Venezuela.

Hooker did much of the initial work on digestive systems of carnivorous plants, and his statement on digestion can be considered an overall definition for carnivorous plants. He said of *Nepenthes*:

> *It would appear probable that a substance, acting as a pepsin does, is given off from the inner wall of the pitcher, but chiefly after placing the animal matter in the acid fluid.*

Hooker is making a crucial point here, that the plant responds to the presence of prey.

On one of Hooker's journeys to India he used glass cases

to transport plants back to Kew. Some of these cases were so heavy that they weighed up to 174 kg (three hundredweight) and made it possible to bring back whole trees, as well as rare and highly sought after species of *Nepenthes*, orchids and other tropical plants.

Hooker died in 1911 at the age of 95 leaving a legacy of the largest collection of living plants to the world. William Thiselton Dyer, son-in-law to Hooker, took over Kew Gardens in 1885 and added a special glasshouse to house his father-in-law's favourite plants, the *Nepenthes* pitcher plants. To this day the hybrid *Nepenthes hookeriana* carries his name.

That Hooker was able to use these glass cases is due, in part, to the work of Nathaniel Ward. During long sea voyages plants would often dry out or die from being watered with sea water. Prior to 1830 all *Nepenthes* introduced from the East were lost. In 1830 Nathaniel Ward discovered that plants in sealed glass cases could survive many months, even years, if given sufficient sunlight. These cases became known as 'Wardian Cases', and were made from glass with wood or lead edges. Each had its own self-contained environment. These cases are still in use today and are more commonly called terrariums, and are still a good way to grow many carnivorous plants.

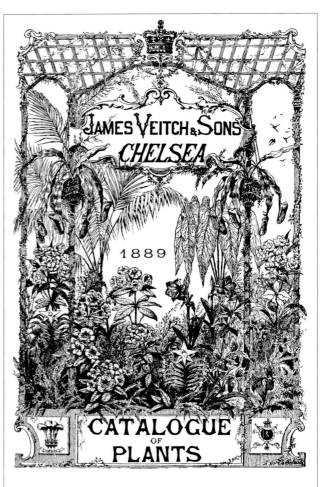

Some of the species in Veitch's catalogue were quite well described and many were accompanied by a detailed engraving, originally carved on wood.

Wardian cases, like this one illustrated in James Veitch's catalogue, were invented in the 1830s and made it possible for plants to survive in sealed cases for many months.

Ward's invention made it possible to import more and more plants into Europe from the tropics, with very few dying on the long journey. This was a breakthrough for botanists, who could now work on live specimens instead of the dried plants of the past—they could actually see if plants were indeed carnivorous. Also, for the well-to-do Europeans, palms and wardian cases became a feature of home decor. Stove houses (or heated glasshouses) became popular, as did insect-eating plants, especially the exotic *Nepenthes* from the tropics.

So widespread was *Nepenthes*' popularity that major nurseries such as that of James Veitch of Chelsea, London, sent scouts all over the world collecting these and other rarities such as palms, orchids, and hoyas. Many species were named after the collectors that worked for Veitch, for example *N. burbidgeae*, named after the wife of F.W. Burbidge; *N. burkei*, after David Burke, and *N. veitchii* after the owner himself. His nursery also prided itself on developing some excellent hybrids, once again named after the hybridiser: *N. dominii*, the first artificial hybrid *Nepenthes* ever raised, named after John Dominy, and *N. courtii*, named after William Court. James Veitch said that the majority of hybrids raised under artificial conditions proved more amenable to cultivation and were superior to their parents, resulting in a stronger plant.

The plants were detailed in a catalogue. Its 75 or more pages also covered palms, orchids and other species. The range of carnivorous plants was extensive for the 1880s, and included, at various times, the Northern American Pitcher Plant (*Sarracenia*, including hybrids such as *S. courtii*), Bladderworts (*Utricularia*), the Sun Pitcher (*Heliamphora*), the Albany Pitcher Plant (*Cephalotus follicularis*), Cobra Lily (*Darlingtonia californica*), Venus Fly Trap (*Dionaea muscipula*) and various species of Sundew (*Drosera*). Prices ranged from three shillings and sixpence to twenty-one shillings for the rare plants. Some exceedingly rare species could fetch as much as one hundred and five shillings. These plants were very expensive, and only available to the upper class. Within this class no stove house was considered complete unless it had a *Nepenthes* dangling from the rafters. *Nepenthes* could be brought inside the home from time to time, as a conversation piece, much to the delight of visitors, in the same way as we might bring a Bonsai indoors today.

James Veitch sent some of the collected plants to Dr J.M. Macfarlane, professor of Pennsylvania University, USA, who, with the help of Kew Gardens and other botanists, wrote *Observation on Pitchered Insectivorous Plants*. The first part appeared in 1889, the second four years later—this work is still considered a milestone. Macfarlane's work covered pitcher plants—*Nepenthes*, *Heliamphora*, *Sarracenia* and *Darlingtonia californica*. Unlike Darwin, Macfarlane attempted to explain the evolution of each species, suggesting that all these pitcher plants originated from one common ancestor.

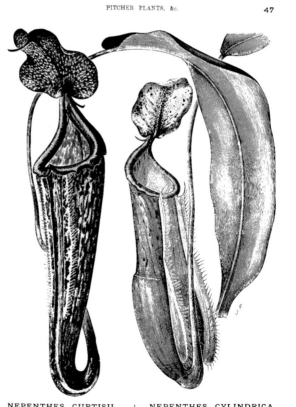

These pages from Veitch's catalogue indicate both the scientific and popular interest in carnivorous plants in the nineteenth century, although palms, orchids and other 'exotic' species were also illustrated. Nepenthes, *such as those shown here, were particularly popular: they were also, as is quite clear from the prices listed, available only to the wealthy classes.*

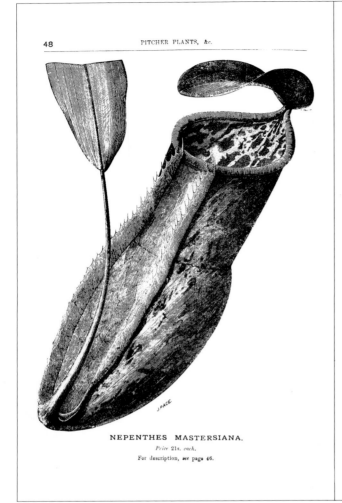

James Veitch's nursery not only collected plants from all over the world, it also developed its own hybrids, such as Nepenthes rufescens, *described in the pages of his catalogue. Notice the species* N. veitchii, *named after the nursery owner himself.*

Macfarlane was the first to suggest that pitchers of *Nepenthes* could be trimorphic, that is having up to three different pitcher shapes on the one plant. Much of Macfarlane's work was on cell structure and very detailed, and while some of his conclusions have proved to be incorrect, many were correct. Macfarlane's studies of *Sarracenia* revealed that secretory glands existed on the outer surface of the walls, to lure insects up the walls of the pitcher to the inside. I often wonder, given what these botanists did achieve, what Darwin and Macfarlane would achieve today with the aid of high-powered electron microscopes!

A more recent contribution to the work on carnivorous plants occurred in 1942, when Volume IX of *A New Series of Plant Science Books* was published. It was written by Francis Ernest Lloyd, and was later renamed *The Carnivorous Plants*. This book covered all major studies by other botanists on carnivorous plants, as well as detailing work by Lloyd himself. Lloyd was Professor of Botany at Magill University in the USA. He died in 1947 at the age of 79, having devoted much of his life to the study of carnivorous plants. Lloyd's work covered many crucial issues, but unfortunately I can mention only a few.

The evolution of carnivorous plants fascinated Lloyd, but he came to no definite conclusion on it. He was hampered by the fact that 'fossil evidence is meagre' and any theory and conclusions tenuous. He points out that while some plants are rare, existing in only one or two places and therefore indicating their possible isolation from the rest of the world, other carnivorous plants are widespread. In some areas, however, both the rare, supposedly isolated, species and those widespread exist side by side. Hampered by these possible contradictions, aware of the problems they raised, evolution of these plants remained a mystery to Lloyd. Even so, he went on to examine carnivorous plants, concentrating on their trapping mechanisms.

For many years after Lloyd's contribution, it appeared that little more could be said about carnivorous plants. Most of the work simply refined his theories using the latest technology, and redefining some of the misnamed genera. Recent discoveries of the carnivorous habit of *Triphyophyllum*, however, and possibilities for carnivorous vegetables, indicate that there is still a lot to be learnt about the plants of this world. There are still areas that may, in the future, reveal new carnivorous plant species. For the present it is enough to examine some of the 550-odd plants that we *do* know are carnivorous, to postulate as to how they might have evolved.

'Normal' Plant to Carnivorous plant

How did carnivorous plants evolve in such a diverse way? How is it that today hundreds of species exist in a range of different climates and soil types, in almost every country in the world? How did such varied trapping mechanisms—spring, suction, sticky leafed and pitfall traps—develop?

A Victorian hybrid of Nepenthes maxima *and* N. x superba. *The three pitchers on the right are winged lower pitchers while the fourth (facing away) is an upper pitcher.*

Drosera villosa

Fossil evidence for links between carnivorous and 'normal' plants does not exist, so the path or paths carnivorous plants took can only be conjectured. What we do know is that *all* plants have movement, usually slow and slight, such as leaves tilting towards sunlight. Most plants have the ability to absorb nourishment through leaves, and roots secrete a liquid to aid movement through the soil. Many plants hold insects to aid pollination. When a plant is able to put together these factors, and actually trigger a response to its prey, it is called carnivorous.

There are essentially two schools of thought on the evolution of carnivorous plants. One suggests that, as plants evolved, some were able to consolidate all these elements and produce a variety of different-looking carnivorous plants, in different parts of the world. The other school suggests there were one or more plants with many features (some of which might not even exist today) that were lost or modified to develop into the variety of carnivorous plants we know today.

Without fossil evidence to resolve this debate, it is instructive to look at the mutations that occur occasionally in carnivorous plants. When stressed, most carnivorous plants *lose* their carnivorous plant characteristics rather than increase them, in order to save the plant and almost as a retaliation against environmental pressures.

Some strange leaf structures develop when plants are under stress as a result of environmental changes. The *Sarracenia* pitcher, for example, develops into a flat 'normal' leaf during dry conditions. *Nepenthes*, when stressed as a result of low humidity, produce either no pitcher or just a hook, and when overfertilised these plants have no need to be carnivorous, and hence produce no pitchers. Other strange leaf forms can arise, including a series of spikes on the leaves of *Dionaea*, and a half leaf–half pitcher on *Cephalotus* (reminiscent of a developing *Nepenthes* leaf). Apart from environmental responses and possible mutations, factors to consider when trying to map an evolutionary path are the spread of plant species throughout the world, the shape of their floral parts, the structure below ground level, and the cell structure that make up the carnivorous characteristics.

Considering all these factors, I have developed what I feel are two possible paths to 'carnivory'. No plant more represents the transition from 'normal' plants to carnivorous plants than *Triphyophyllum peltatum*, looking as it does something between a passionfruit, a *Nepenthes* and a *Drosera*. It provides three different leaves in response to different signals. *Drosera* may have evolved from some primitive form of *Triphyophyllum peltatum*, as *Drosera* has such a wide variety of leaf shapes, and is one of the widest spread genera. *Drosera* also has all the characteristics necessary for a gene pool from which all other carnivorous plants could have evolved.

It may have been possible for any of the *Drosera*, *Nepenthes*, *Byblis*, *Pinguicula* or *Heliamphora* genera to have developed independently. On balance I feel this is unlikely.

One last, highly speculative, option is that, at some time in the past, there was a 'super carnivor', a plant with all of the features of the above plants and maybe more we couldn't dream of. It may have existed all over the world and, over time and with evolutionary pressures, lost many of its features, changed others, and evolved into the diversity we know today.

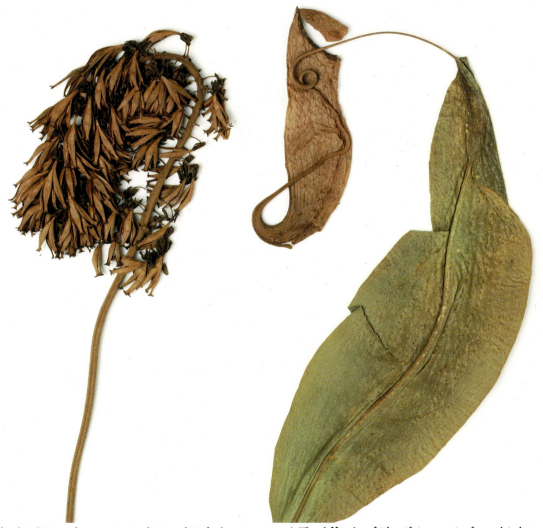

This is the dried Nepenthes *specimen discussed in the letter on page 4. The difficulty of identifying species from dried specimens is clear—colour, texture and shape are hard to specify. Further, different stages of a single plant, when presented as specimens, were easily attributed as separate species.*

EVOLUTIONARY PATHS

Here are two possible evolutionary paths, from non-carnivorous to carnivorous plants, as discussed in previous pages. Both are theoretical paths, including both actual and imagined plants. The labelled plants are the same ones used to illustrate each genus in the descriptive section of this book, and the other stages are possible intermediate plants. These intermediate plants are not reproductions of plants, but serve rather to show significant features that have changed, developed, *evolved* into a state of carnivory.

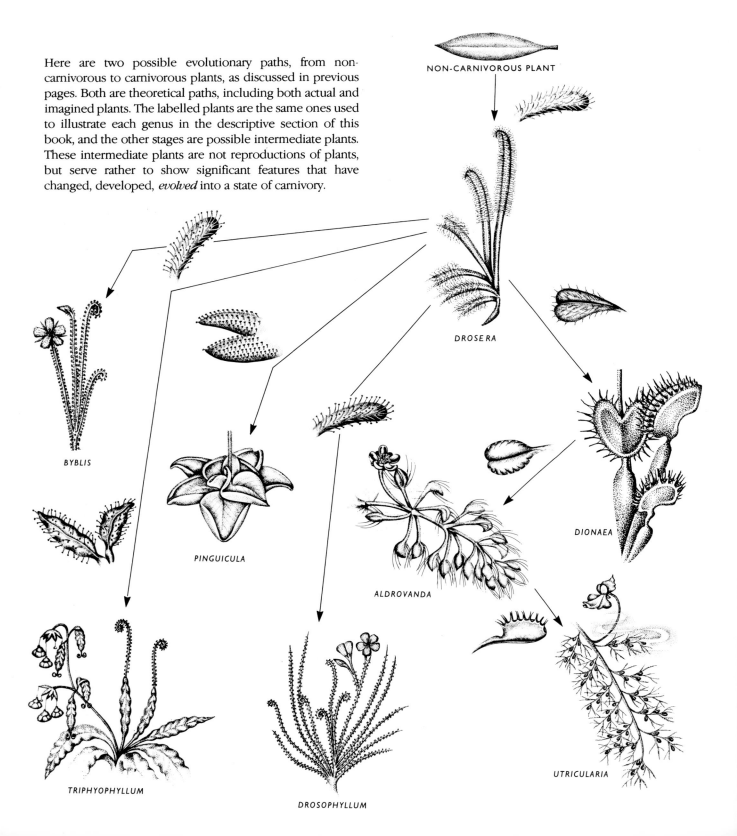

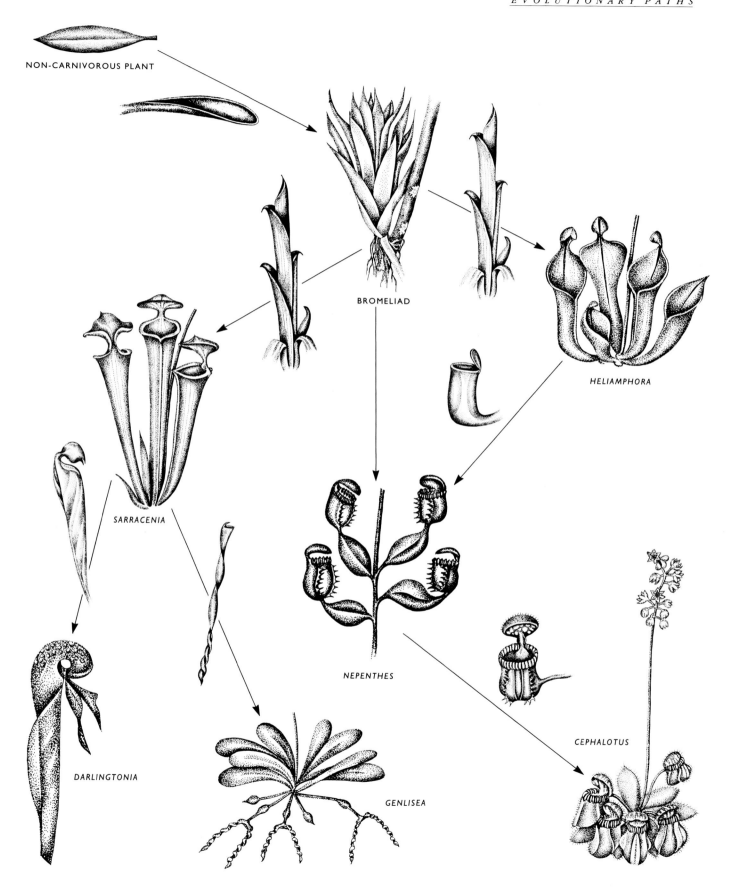

Parts of Carnivorous Plants

Carnivorous plants are varied, and there is no *one* 'typical' plant that can be divided into parts as a representative of all species. Instead, here are some interesting and attractive features of carnivorous plants—from flower to pitcher to leaf, which are also discussed in the Genera, Trapping, Cultivation and Propagation sections.

Nepenthes villosa: *has wings on the pitcher and a spur at the top of the lid.*

Dionaea muscipula: *has trigger hairs on the hinged leaf-blade traps.*

Sarracenia purpurea *ssp* venosa: *has a ruffled hood on the pitcher.*

Catopsis berteroniana: *an epiphyte (grows high on trees, without being a parasite).*

Heliamphora heterodoxa: *has a red hood, and a red rim that defines the edge of the pitcher.*

Ultricularia vulgaris: *the bladder (with airbubble) allows this aquatic species to stay afloat.*

PARTS OF CARNIVOROUS PLANTS

Nepenthes masoalensis: *its red glands secrete a sugary substance to attract insects.*

Drosera filiformis: *the sticky tentacles on the erect stem bend toward the prey.*

Drosera peltata: *peltate (cup-shaped) leaf, with red tentacles, on a fine petiole.*

Sarracenia oreophila: *showing phyllodia (winter leaves).*

Nepenthes lowii: *has sugary secretions on lid, and juvenile pitcher in background.*

Nepenthes rajah: *has fringed wings, a spur and closed pitcher lid.*

Drosera paleacea: *the flower has five petals and five stamens.*

Sarracenia flava: *flowers are noddling.*

Nepenthes: *female flower with seed capsules.*

Sarracenia rubra: *this flower has two short pale stigma points, one on either side.*

Cephalotus follicularis: *note the new white secondary rhizome and the old reddish brown rhizome.*

Nepenthes: *this male plant has a branched inflorescence and one-flowered pedicels.*

Classification of Carnivorous Plants

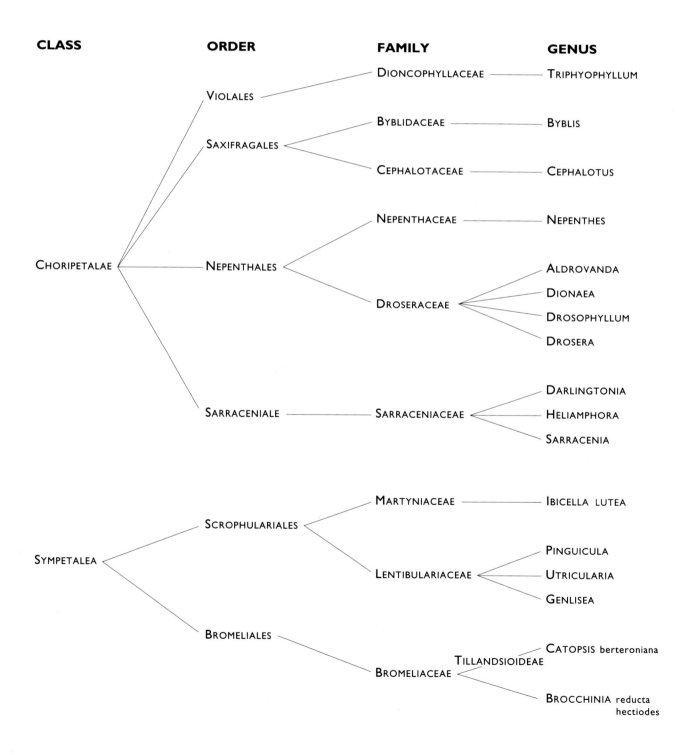

Drosera rotundifolia

Trapping Mechanisms

Trapping mechanisms of carnivorous plants are both fascinating and diverse. It is impossible to make generalisations about them without speaking of qualifications, exceptions and changes. As a result, I have decided to look at *specific* trapping mechanisms, as a way of exploring both the similarities and differences between carnivorous plants.

Cut away shot of the trap of Sarracenia purpurea: *note slippery sides.*

Cut away shot of the trap of Cephalotus follicularis: *note curved teeth and rim.*

Dionaea muscipula: *note that trap is full of fluid.*

Cut away shot of the trap of Sarracenia leucophylla.

Cut away shot of Sarracenia psittacina: *note that the hairs all face one direction at the entrance.*

Dionaea and Aldrovanda

Dionaea and *Aldrovanda* both have spring traps, that spring shut once an insect touches its trigger hairs.

On either side of each lobe are three sensitive trigger hairs, 4 mm–8 mm (0.15–0.3 in) long which, when touched, spring the trap into a semi-closed position. In this state the marginal bristles bend further inwards forming a cage containing the trapped insect. Many believe the trap springs shut at the 'hinge', but this is not so; instead the trap closes about half way up the lobe. One way to see this is to cut down the hinge of an open trap and remove one of the lobes, then touch the trigger hairs of the remaining lobe and see where the trap bends inwards. As trigger hairs on any half-closed trap are further stimulated, the trap closes even tighter. Erasmus Darwin, John Ellis and others believed that the trigger hairs actually pierced the body of an insect, and the lobes crushed it to death. In reality the fine trigger hairs are well hinged and simply bend until they lie parallel with the lobes, in a closed trap. Once the trap is fully closed the surface glands of the lobes exude a liquid consisting essentially of three main ingredients—water, chitinase and proteolytic enzymes. The water helps drown or suffocate the insect, chitinase enzymes help break down the insect's exoskeleton, and finally the proteolytic enzymes break down the proteins. Depending on the type of insect, the plant actually changes the proportions of this liquid. This was indicated by a number of experiments, where increased

Dionaea muscipula: *rare red form.*

amounts of uric acid on the lobes resulted in larger amounts of protein, and hence digestive enzyme, released. An application of ammonia, on the other hand, resulted in less enzyme released. The significance of this is that uric acid is excreted by flies when trapped. The secretion is vital to the plant, keeping the trap closed when insect movement, and therefore trigger hair stimulation, has stopped. After less than eight hours insects have died and it is their bodily fluids that complete the process and keep the trap closed.

Dionaea muscipula

Dionaea muscipula *0.05 seconds later.*

When these body fluids eventually stop flowing the trap begins the slow process of opening. As the soft parts of any insect are now dissolved, it simply remains for the wind to blow away any skeletal remains.

If plants are growing in less than favourable conditions, (low light, cold temperatures, low humidity) the trap closes slowly. Under very poor conditions it may not close at all. This plant is so affected by cool temperatures that when first grown in northern Europe the traps closed so slowly that few believed the plant could realistically catch any insects—anyone suggesting otherwise was ridiculed.

Dionaea muscipula

Exactly how the trap closes is largely unknown, as is precisely how it opens. What *is* known is that electrical activity (action potential) is sent from one leaf lobe to the other upon stimulation of any trigger hair. If this did not happen the lobes would not close together. (Action potential is also the term used to define electrical activity associated with muscle contraction in animals.) Each lobe has an inner and outer surface—once stimulated the cells of the inner surface contract, while the outer surface expands by up to 30 per cent, bending the lobe inward. The outer surface feels quite soft before closure, compared to its state when the trap is in tension and closed. Once the insect has dissolved the cells of the inner wall expand up to 15 per cent of their original size while the outer walls remain constant, causing the lobes to bend outwards. These expansions are irreversible. This system of rapid cell expansion may have consequences for rapid growth in 'normal' plants.

While the precise method of cell expansion and secretion of enzymes is unknown, a few facts are known. The plant has some form of simple 'memory', possibly related to electrical build-up. To close a trap under 'normal' conditions takes two stimulations of the trigger hair, provided they are no more than 25 seconds apart. If the contact is as long as 20 minutes apart, not two touches but 14 are needed to close the trap.

If you have ever opened a trap that has been closed for one day or more you will see an insect awash in an enormous amount of fluid. Any insect that has not been crushed by the two lobes snapping at lightning speed (up to 0.03 seconds) has probably drowned. In very few instances I have seen a small insect escape from a trap by eating a hole to freedom near the hinge. The *Dionaea* trap is so strong and sudden that insects are often trapped half in and half out, between the caged walls. As the trap is so highly-tuned, catching an insect only half way may not release the plant's enzymes and the whole leaf can turn black. If this happens to all the traps, the whole plant can rot away.

Traps die after three to four insects have been caught and dissolved in them, as if the excess food and growth has prematurely aged the leaf. Depending on the age of the plant, size of insect and environmental conditions, a leaf that dies is normally replaced by two or more new leaves, and this process should be considered part of the plant's natural cycle. Problems can arise when traps are continually triggered shut (without any food gain) in a short period of time. If this continues not only the traps but eventually the whole plant will die.

CEPHALOTUS

It was originally thought that the lid on the pitcher of this genus closed as the prey approached. The only time the lid closes is when the plant is not well, or as a safety precaution if humidity decreases. The open lid, with its translucent sections, appears to work like a mirror maze, reflecting the light on the surface of the water-filled pitcher below. The insect thus becomes confused and either flies into or walks toward the reflected opening. An insect is more likely to enter an area that is lit than one that is dark. The translucent section of the lid also allows light to the base of the pitcher, which may aid in the growth of bacteria that helps break down the prey. Digestion increases as temperature increases, and below 10°C (50°F) it stops. This may be why during winter only leaves (not pitchers) are produced. Both the rim and lid look like glistening flesh to an insect flying overhead. Any insect entering the pitcher (even with suction pads) and crawling over the rim in search

Cephalotus follicularis

of nectar is faced with a slippery surface on which it is impossible to get a hold. Further, the rim is shaped like curved teeth overhanging the entry, preventing any escape. Eventually the insect gives up the struggle, falls into the liquid and is dissolved by bacteria and the digestive juices the pitcher exudes. New pitchers contain digestive liquid even before the lid is open, so the plant is ready to catch prey just as soon as the lid breaks free from the rim.

Darlingtonia

Darlingtonia is a passive genus, that is the plant does not move in response to an insect's approach. Instead, it has a confusing window-maze together with a twisted tunnel of downward-pointing hairs that guarantees the capture of any intruder. In juvenile pitchers, before the development of fangs, trapping does not usually occur. In young pitchers, flying insects are initially attracted to the fleshy colour of the fangs. Having arrived on the fangs of a mature pitcher, the insect can sample the nectar exuded from glands. It then samples even more by moving inside the pitcher to feed on nectar produced just inside the rim. Once inside, flying insects spend the rest of their life trying to find the entrance. Crawling insects are also catered for, and as they pass the nectar they enter a waxy zone that continually gives way, and has the effect of scaling a roof with all the tiles coming loose and falling off. The only other accessible path leads down the pitcher, past the downward pointing hairs, to the base. There, the prey is dissolved by digestive enzymes and bacteria. From the downward path there is no return. The only exception is when the pitcher is young and the insect can eat a hole through the pitcher. I have only seen this happen with a very new pitcher, newly opened, with no build-up of insects. Many a European wasp has been discovered dead with its head pointing through a hole in the pitcher wall, unable to get the rest of its body through. In older pitchers, where nectar has ceased to appear on the fangs, the build-up of decaying bodies produces a scent that is sufficient attraction for any insect in itself. Where insects have built up, the fumes from this decomposing 'soup' seem to knock out most intruders within a few hours. You can observe this when a pitcher is full to the brim with insects, where the last few are lying on top, stunned but not trapped by any hairs. Slice down the pitcher with a knife and, within an hour, the last few insects move around then fly away. The range of insects caught includes ants and earwigs and, as *Darlingtonia* grows larger, flies, wasps and mosquitoes and, under cultivation, includes slaters and slugs. Even with the harshest downpours the shape of the head makes it impossible for the prey to wash away.

Where conditions are poor or have changed adversely, new pitchers turn black at the top and traps do not form until conditions improve. When older pitchers have been growing happily and humidity drops, the fangs turn brown, perhaps as a way of preventing build up of insects that it lacks the resources to digest.

Genlisea

Genlisea grows in swampy areas, often with a changing water level that can completely cover it so leaves are hidden with only the flower appearing above the water level. The flower is similar to, and often grows alongside, *Utricularia*. Also similar to *Utricularia*, the trap is white, but instead of a bladder the trap looks like the letter 'Y' and measures up to 37 cm (15 in) in length. The trap is best described

Genlisea hispidula: *underground trapping apparatus.*

as two twisted ribbons open at the ends and along the seams, with one of the ribbons twisting clockwise, the other counter-clockwise. Each ribbon can measure up to 25 cm (10 in), long and they join together to form a bulge or bulb—it is the bulb area that insects are finally trapped, and digested with the aid of digestive enzymes. The rhizome is 25 cm (10 in) along from the bulb, connecting the cluster of leaves and the flower scape. An immature trap simply has a bulb and either no arms or two short ones.

Sarracenia purpurea x rubra

SARRACENIA

The *Sarracenia* pitcher has nectar-producing glands on the outside, and also produces nectar on the rim. This entices the insect down the pitcher, past the rim, to the digestive zone. For crawling insects the top half of the pitcher produces a waxy lining, similar to that for *Darlingtonia*, and flying insects become so confined in the tunnel towards the base that flight is impossible.

Essentially each pitcher has five main zones, as follows:
1 The entrance: The inner lid, containing nectar glands and strong downward pointing hairs.
2 The feed: The smooth surface of the rim and inner lip contains nectar. This section is about 30 per cent of the entire pitcher.
3 The trapfeed: The inner pitcher contains waxy walls interspersed with digestive glands. These come in handy as the insect bodies reach toward the top of the pitcher. This area takes up about 45 per cent of the pitcher.
4 The blocked path and absorption: This lower, deadly, section of the pitcher contains downward-pointing hairs, glands that produce digestive enzymes, and those that absorb nutrients. This area takes up about 10 per cent of the length of the pitcher.
5 Food canal: This final zone supports the tube and transports nutrients to the rhizome.

The nine different species of *Sarracenia* have varying length zones, but essentially retain the proportions described above.

Sarracenia flava: *pitcher cut away to show build up of insects.*

UTRICULARIA

Utricularia traps are tiny—often no larger than a pin head—and very complex, involving a mixture of attraction and direct stimulation. Stalked mucilage glands on the 'door' and on the cells beside it contain sugar that attract the prey that branched antennae have already lead in. Two or more trigger hairs, once stimulated, cause the opening of this aquatic trap to swing inward in as little as 0.02 seconds: as the inside of the trap is a vacuum, it sucks in any nearby prey. The sugary mucilage on the door seals it tight. The walls of the trap are only two cells thick, and they bend outwards as they secrete excess water, and the prey is dissolved in anything from 75 minutes to 20 days. If a trap has caught a large insect it will eventually suck the body in sections. Full traps turn black and fill with waste material—a plant usually has a number of full traps and a number primed for trapping all at the same time. The size and shape of these traps differs between species, but this is the basic trapping principal for all *Utricularia*.

Sarracenia leucophylla

Cultivation

Although carnivorous plants exist naturally in many different climates, most can be grown together in similar conditions if your climate is temperate to subtropical. In particular *Dionaea*, *Sarracenia*, tropical *Drosera* (except tuberous), *Pinguicula* (except cool species), many *Utricularia* (except tuberous) and *Genlisea* can all be grown in similar conditions. The rate of success when growing carnivorous plants usually depends on the degree of variation between the original environment of the plant and that of the grower. Nevertheless, it is difficult to make generalisations about some genera, as so few have ever been cultivated, while others—such as *Triphyophyllum*—are difficult to grow even after years of patience. Other carnivorous plants, such as *Byblis gigantea*, can die unexpectedly after years of perfect growth. To combat these difficulties I have moved from general issues such as water and light requirements on to specific examples. Finally, there is a table of Cultivation Guidelines at the end of this book, to bring it all together. Further information on the growth and maintenance of carnivorous plants appears in the Propagation chapter and in the Genera sections.

Growing Medium

As a basic guide combinations of peat moss, sphagnum moss, sand and/or vermiculite can be used as growing mediums for most species. Leaf mould, scoria, chipped bark, perlite, and charcoal can also enhance a soil mix.

Once you have decided on your soil, you must decide what type and size of pot to use. Many commercial nurseries grow carnivorous plants in 5 cm (2 in) black plastic propagating pots, which have a depth of 7 cm (2.75 in) and an opening of 5 cm (2 in) with slightly tapered sides and a smaller base. These pots sit in large trays of water receiving full sunlight for the first few years. After about three years these plants would be potted up into 10 cm (4 in) plastic or terracotta pots. Of course, there are many different containers that the hobbyist can use to grow *Sarracenia*, *Drosera*, *Dionaea*, *Pinguicula*, *Utricularia* and *Genlisea*. These include terrariums, 'water wells' and bonsai dishes in mixed arrangements. These plants can also be grown in glasshouses in hanging baskets, in mixed bench arrangements or outside in artificial bogs, dams and ponds.

When repotting be warned that very dry peat may take several weeks to become fully damp. To avoid this let the peat sit in a bucket of water until it sinks, pour the excess water off and use the soggy peat to fill your container.

Watering Methods

As carnivorous plants can absorb more than five times the amount of water a 'normal' indoor plant might use any contaminants build up quickly, killing the plant. For this reason it is always wise, at least initially, to use rain water or purified water. If you are unsure of your local water supply many water authorities will test your water on request, although it may be too late if you observe a crusty build-up at the base of pots and the surface of soil.

If you grow your plants in shallow trays of water, then the water should be replenished in the early morning, ideally using slow drippers on a timer that activates as the sun is beginning to appear and warm the benches.

Water in trays become quite warm by the middle of the day, thus increasing the general humidity, and warming the soil and roots, so that by nightfall densely packed pots still have quite warm soil. Replenishing the empty trays the next day also prevents any build up of stagnant water. On very hot days, when the temperature exceeds 30°C (87°F), all the water can evaporate, small pots can completely dry out and, after two days of hot weather with no water, leaves droop then wither and your plant might die. If discovered by the end of the first day, the plant could be revived by placing the pot in a large bowl of water under a bench, in filtered light, for one day. Return the plant to its original position the following day, and as a precaution—since plants are more prone to attack after stress—spray with both a general fungicide and a soil fungicide.

An artificial peat bog.

A Sarracenia *showing clear signs of lack of water.*

The bend in this Sarracenia *indicates that the plant has suffered from lack of water, followed by a sudden increase in water.*

Different growers have different methods and can use vastly different water levels. What all good growers have in common is that their water level is *constant*, just as it is in swamps and bogs. The plant adapts with roots that either lie at the surface or travel to the base of the pot for water. It is important to maintain a constant water level for four months over summer and, at a lower level, four months in winter. If you do not do this plants can become so stressed that they may lapse into an early dormancy.

Light Requirements

Light is an important variable in growing carnivorous plants as it affects (among other things) the colour of the plant. The connection is not straightforward, however—excess light can be countered with high humidity and low temperatures to prevent high leaf-surface temperatures.

Light requirements vary greatly. Maximum sunlight as experienced by plants that grow naturally in an open plain, for example, will often ensure full-leaf colour. Some short plants, on the other hand, especially those native to areas with tall scrub, require filtered light provided by surrounding plants, shadecloth, whitewashed glass or placement on a window sill. Still others require full shade, provided by placing plants under a bench, in a fernery or indoors, and these variations will be indicated as they apply.

Drosera helodes

Seeds

Not all species need be cross-pollinated for fertilisation to occur. To fertilise, brush the pollen from one flower onto the stigma of another. The best time to collect seeds is after the flower stem has begun to turn black, usually about five weeks after fertilisation. Some seeds are quite hard (eg *Dionaea* and *Sarracenia*) so simply rub the spent flowers between your fingers for the seeds to fall into an open envelope. Other seeds, such as *Nepenthes*, are soft and fragile, and best collected in a paper or plastic bag covering seed capsules. As winter approaches the envelope of seeds should be placed in the refrigerator for three months so they experience winter dormancy. Assuming your parent plant also experiences winter dormancy then you could simply sprinkle seeds on the top of the soil and wait for the emergence of spring.

CULTIVATION

Fire can be both constructive and destructive assisting in the germination of some species: Anglesea, Victoria

Three weeks later, and the first species are growing back. Note the recuperative qualities of the bush.

Six months later, and reconstruction is all but complete.

Seeds grown in a tray should be sprinkled onto damp peat moss then well watered. After spraying the tray with fungicide, cover it with a plastic bag or place in a closed terrarium. Position it to receive filtered light at a temperature of about 25°C (77°F). Seedlings should appear in about four weeks. When they are about 19 mm (0.75 in) wide they can be potted up. When left too long in the seedling tray, however, seedlings tend to produce less vigorous growth. Repotted seedlings, in contrast, usually display greater growth.

It is important to use sterile soil for growing seedlings. One way of achieving this is to place your tray of peat moss in a domestic microwave oven for a few minutes—allow the peat to cool then plant seeds.

An interesting and enjoyable way to start off seeds is in a Petri dish with a fine layer of peat moss. Spray the dish with a leaf fungicide. Cover with a lid and keep in a warm position with filtered light. Shoots will appear within a couple of weeks: plant up one week later and keep protected from excess light and heat.

DORMANCY

During winter many carnivorous plants need to remain dormant, at temperatures as low as 5°C (41°F). Exceptions are listed in the cultivation table. Maintaining warm temperatures during winter could cause rhizomes to rot. Where it is difficult to achieve temperatures below 5°C (41°F) simply remove the rhizome from the soil and place it in the refrigerator for three months. Before doing this spray the rhizome with a leaf fungicide, divide the rhizome if required, and remove all the leaves. As spring approaches repot the rhizome in new peat moss, water in a soil fungicide, then continue watering till the soil is slightly damp.

Utricularia quelchii

Pinguicula rotundifolia: coming out of winter dormancy, with new summer leaves on the right.

EXAMPLES OF CULTIVATION METHODS

Byblis liniflora

Byblis liniflora can grow in a warm glasshouse alongside *Nepenthes* and orchids. I know of one commercial grower of *Cattleya* orchids who always has a few plants near the entrance to the glasshouse to trap any insects entering, and has thereby reduced insect infestations.

This very fine powdery seed is easily raised when sown during the beginning of spring on damp peat moss. Tiny leaves with droplets will appear, looking like grass—they should be thinned out before a third leaf appears. The seedlings should be planted up into 5 cm (2 in) tube pots and placed on watering trays in an area receiving filtered light. From planting till flowering takes about three to four months. *B. liniflora* dies off each winter and this method of cultivation is so successful there is little point in trying others.

The soil mix for *Byblis liniflora* should be two parts peat moss and one part sand in a fairly tall pot watered by tray. Maintain humidity at about 50 per cent and temperature above 15°C (59°F), and it should keep growing year after year. In its natural habitat it dies off each year during the dry summer. Decrease watering at the beginning of winter, and trim the plant back each second year at the end of winter. In temperate climates a soggy soil will lead to root rot, whereas in subtropical and tropical climates very damp soil is not a problem. *B. liniflora* prefers shade, which is provided by tall grass in its natural habitat.

Byblis gigantea

Excellent results have been obtained with these oblong black seeds (produced in late summer) by letting them experience fire to stimulate germination. To begin, then, sow the seeds on a soil mix of equal parts peat and perlite, or one peat and four of sand. Cover the seeds with about 4 cm (1.5 in) of soil, then cover that with about 20 dry gum leaves and set fire to them. This stimulates the regular fires of their native habitat. One month later shoots will appear, and within six months a new flower-producing plant will have developed.

As an alternative to fire, good results have been achieved by pouring boiling water over the seeds, then planting them. While the strike rate is reduced it is satisfactory to plant seeds direct onto soil.

Cuttings can also be obtained from the base of the plant. Ensure that 2.5 cm (1 in) of the older base is attached, then plant into the soil, and roots will appear in a couple of weeks. This is best done during early spring.

Soil appropriate for *Byblis gigantea* is 'open', using either half peat half sand, half scoria half peat, or half perlite and half peat. The pot needs to be deeper than it is wide to allow the long roots to develop. Each year—or at the least every second year—the plant needs to be cut back level to 2.5 cm (1 in) above the ground. Cut plants back in the following year and three or more plants will begin to sprout for every one original plant: split these up in a month or so. Grow *B. gigantea* in full sunlight, with 50–60 per cent humidity. Do not sit the plants in water as they tend to rot, and water a soil fungicide through every month. Remove the tips of any leaves that show signs of fungus, provide ample ventilation and use a six-month slow-release fertiliser.

Cephalotus follicularis

Planting seeds as a method of cultivation is a slow process. It may be four or five years before a plant is old enough to flower. Methods other than cross-pollination have a germination rate of about 50 per cent, while cross-pollination increases this to 75 per cent. Cross-pollination is easy as, at any one time, there are usually two or more flowers open. As the flowers open the stigma is hardly discernible but the pollen-producing stamens are clearly visible. There

Byblis gigantea

Cephalotus follicularis

is a time lag between pollen production and the stigma being receptive. After a few days the pollen ceases to develop and the stigma is visible, and receptive to the pollen. It is best to collect the pollen using a small paint brush, storing it in a small vial in the refrigerator for a few days till the stigma is visible, then brushing the pollen on the stigma. After a week the swollen ovaries indicate success. Six weeks later fine brown seeds will develop, and they need to be kept cool in a refrigerator for two or three months before planting out. Seeds are best placed in a polystyrene (foam) box with sides of about 15 cm (6 in) and drainage holes. The soil should be made of three parts peat moss and one of sand—this will provide ample drainage and prevent seedlings from rotting. As a precaution it is best to spray with a fungicide every three months and water through a soil fungicide every two months. By mid-winter a mass of small leaves cover the box, then some 12 months later, the soil level having dropped 5 cm (2 in), the seedlings are ready to be potted up into 5 cm (2 in) propagating pots containing peat moss only.

Water from overhead or, if you need to water by tray use a soil mix of three parts peat and one part scoria. Once the pitchers and leaves are overhanging the pot, and roots are beginning to appear from the bottom, pot the seedling into 10 cm (4 in) pots. Once outgrown, pot into a hanging basket, bog garden, terrarium, 30 cm (12 in) tub or mixed arrangement. Line the hanging basket with sphagnum moss, and use either of the soil mixes mentioned or, if you decide to grow your plant in a shallow bonsai dish, use peat moss and place the pot on a bed of damp sphagnum to prevent drying out or propagate via leaf and pitcher cuttings. For details of this method, see the Propagation chapter.

For best results grow this plant in full sunlight: the rim and lid will turn burgundy, becoming darker and covering the entire pitcher as sunlight increases. Cool nights, where the temperature drops to below 10°C (50°F), also produce dark coloured pitchers. In its natural habitat the temperature can rise as high as 40°C (104°F) during summer and drop to 10°C (50°F) in winter. When the temperature increases to 30°C (86°F) and the humidity drops, the lid on *C. follicularis* closes. Also, if the soil is dry the pitchers become soft and will wither unless humidity and soil moisture are adjusted. *C. follicularis* can survive and grow in filtered light, but while the pitchers and leaves will be bright green, flower stems do not usually appear, and those that do, do not often produce pollen. Plants watered by tray in the early morning should absorb the water within a few hours.

Cephalotus is prone to root rot under cultivation (there is a good chance root rot will occur if plants are left sitting in water during winter), so use a soil fungicide every couple of months to ensure good strong growth.

Cephalotus can be grown quite successfully in a terrarium for many years, if the terrarium receives adequate light (perhaps provided by fluorescent tubes), but does not overheat. Also ensure that neither the pitchers nor the leaves touch the sides of the glass, as water running down the glass can rot leaves and pitchers. Very large plants, or clumps, can be planted in 60 cm (24 in) tubs. Place them in 5 cm (2 in) deep trays of water or dig them in beside a watercourse.

A good soil mix to use is an open one, made of three parts peat, one part vermiculite and two parts sand: it is ideal for seedlings in small tube pots, and for mature plants in 10 cm (4 in) pots. Sphagnum moss grown on the surface provides increased humidity. As *Cephalotus* grows close to the coast in Australia, it is one species not affected by salt water. It grows along inlets and creeks, so root systems are kept cool producing healthy plants. For this reason, clay pots or polystyrene boxes make ideal containers.

Dionaea

The best way to care for your *Dionaea* is to duplicate, as much as possible, the plants' natural habitat. This involves providing a moist soil that consists of either just peat moss, or three parts peat and one part sand, or three parts peat and one part vermiculite. I have seen very healthy plants growing in all three mixes. Plants that are repotted in fresh soil and divided each year are healthier than those in the same container year after year. Repot at the end of winter.

The best way to grow *Dionaea* is outside in sunny artificial bogs, or around ponds and dams, where they will produce short leaves and red traps, have large rhizomes and need little, if any, chemical spraying. *Dionaea* grown in a glasshouse under constant fine misting, in contrast, tend to produce long stringy, soft leaves, and have small rhizomes in relation to leaf growth and greener traps. They will also be prone to disease, insects, extremes of temperature and, as a result, will require constant spraying with insecticides and fungicides.

Dionaea muscipula

A plant in poor soil will respond to repotting within a couple of weeks by sending up three or four new leaves, while plants growing in poorly lit areas will become bright red within the trap in less than a month once moved to full sunlight. *Dionaea*, like many other carnivorous plants, can be grown in a variety of ways, in shallow bonsai pots, hanging baskets of sphagnum moss, water wells on sunny window sills, bog gardens, or terrariums lit by artificial light. Commercial nurseries use large well-lit, well-ventilated glasshouses.

As autumn approaches you will notice the tall, erect summer leaves beginning to die, only to be replaced by shorter winter leaves. This is an indication of the onset of dormancy. Eventually all the summer leaves will disappear and only a short rosette that grows flat to the ground will be left. *Dionaea* needs at least four hours of sunlight per day during summer. If a plant receives less than four hours of direct sunlight, it will have very thin leaves and a green trap interior, with a poor ability to dissolve insects, and will shortly die. Remove black or dead leaves as they appear as they are prone to grey mould (*Botrytis*).

Growing *Dionaea* outside in suitable climates can provide the plant with a natural winter dormancy. One of the surest ways to kill *Dionaea* is to maintain high levels of humidity, water and warmth during winter. Under these abnormal conditions the leaves and traps will look quite reasonable, while the rhizome beneath is rotting away. During this winter dormant period, lasting three to four months, plants that were sitting in water should be allowed to drain and should be watered only about once a week. One month or so after the onset of dormancy the rhizome can be divided and repotted in new pots and soil. Remove all green parts of the *Dionaea* during dormancy and, after washing the rhizome, dip it into a solution of fungicide, then place it in the refrigerator for two to three months (not in the freezer as it will turn black and die). This method of artificial dormancy is particularly useful when you are preparing for a show and need to control emergence of leaves and flowers.

Dionaea muscipula *emerging from winter dormancy.*

Darlingtonia

Natural habitats of *Darlingtonia* include mountain ranges with constantly running cool water. Growers in tropical regions have had little success growing this species longer than a couple of seasons. *Darlingtonia* grows best where the root system is kept cool in, say, a clay pot or foam box. Place the containers on an 8 cm (3 in) deep bed of damp sphagnum moss or in large concrete tubs. Otherwise grow *Darlingtonia* beside a freshwater fish pond or dam. Ensure the plant has access to the water level but is not sitting right on the water's edge, achieved by having a base layer of sand, with 25 cm (10 in) of peat moss on top.

The best time to repot or plant near a waterhole is before summer. If it is not possible to ensure a cool root system during summer, however, and the daytime temperature exceeds 25°C (77°F) then about 50 per cent shade is needed, or mist every half hour to cool the plant down and maintain the ideal humidity of 50–70 per cent. If in doubt about temperature and humidity levels, watch the fangs of young pitchers, as they begin to dry, curl up and go brown with continuous heat and low humidity.

Use a slow-release fertiliser with a three to six month life span to encourage flowering, and apply it at the beginning of spring.

The soil mix can be either sphagnum (replaced every year), peat moss (replaced every two years), a mix of equal parts peat and sand, or three parts peat to one of vermiculite. *Darlingtonia* can also be grown hydroponically, using sand or a 'soilless' mix such as perlite or vermiculite. Using an hydroponic method, cool water trickles from hoses buried below the damp ground and keeps the roots cool.

Dormancy should last three to five months, with the daytime temperature below 10°C (50°F) and the nights between 0°C and –5°C (32°F and 23°F).

Drosera

Tropical and subtropical

High humidity and warmth will generally ensure *Drosera* grows larger each year—less than ideal conditions result in a plant that grows like an herbaceous annual. Where shade is limited place plants in trays of water. Most *Drosera* are ideally suited to a terrarium, *Nepenthes* glasshouse, warm window sill or mixed bench arrangement, although *Drosera* may take over. So prolific are the roots of many *Drosera* that they can even grow out of drainage holes in pots in soil-free water depressions on benches. The soil mix should be three parts peat and two parts sand for the subtropical species, and sphagnum moss only for the tropical species.

Tuberous

Plant seeds in early autumn, and keep damp till early summer (I have known of growers who, having planted seed, found that germination had not taken place, and discarded the

Drosera *showing red digestive enzyme.*

pot only to find shoots appearing one year later). In their native habitat, most tuberous *Drosera* revert to tubers during the hot dry summers, thus plants grown in pots should be allowed to dry out totally during summer if grown in a greenhouse, and watered lightly at night if grown outside. Place the pots under benches until the beginning of winter.

As winter begins, start to water these pots when the first signs of life appear.

One tuber planted in summer can result in three or more tubers by the following summer, and in small pots these usually appear sprouting between the side of the pot and the soil. If planting a tuber that has been sent to you try to determine whether the plant has experienced dormancy, and if this is the case, plant in soil in autumn and keep the soil just damp, then, once growth appears, water by tray. If you obtain tubers towards the end of winter, keep the tuber dry and plant up the next autumn—if watered before autumn the plants will begin to shoot earlier. If dormancy is restricted and plants are kept damp all year round, in filtered light, then plants tend to be thinner successively, they will be less healthy and will not produce good growth. After a couple of seasons the tuber then rots away. Use fairly long pots, 10–15 cm (4–6 in) for most tuberous *Drosera*; however with *D. gigantea*, a 30 cm (12 in) deep pot produces best results, planted 2.5 cm (1 in) from the base.

Tuberous *Drosera* do not readily reproduce from leaf cuttings, and small plantlets that survive a couple of months usually die during summer. Further, they do not last long in terrariums, as they tend to grow stringy and soft, and be prone to fungus attacks. Even during winter, the tubers of plants left sitting in water will rot. Water the pots from overhead, and allow to drain. A thin layer of scoria on the surface prevents particles of peat splashing the tentacles when watering. Use a soil mix of three parts peat and two of sand.

Drosophyllum lusitanicum

Plant in three parts sand and one part peat in a clay pot with no drainage hole, and add 1 teaspoon of limestone or dolomite to each 15 cm (6 in) clay pot, or 2 teaspoons to a 30 cm (12 in) pot. Place the clay pot inside a plastic pot containing live sphagnum moss. Ensure the plastic pot *does* have a drainage hole, as it will take up water from a tray. In this way the plant is never watered from above, but absorbs through the clay. Allow the pot to dry out for about one day each month and you will have a plant that grows rapidly for a number of years. *D. lusitanicum* require ample sunlight—less than ample sunlight results in spindly plants—but avoid temperatures above 30°C (86°F) and provide a cool winter. This plant is very difficult to maintain in the tropics and is best grown in temperate or mediterranean climates.

To aid in germination either remove the tip of the seed at the narrow end by snipping it off with scissors or by sandpapering it down, or pour nearly boiling water over the seed and soak it overnight. Seeds have been known to take two years or more to germinate. Plant the seeds on sphagnum moss and cover with a glass jar slightly tilted to provide ventilation. Seeds planted in autumn should germinate six to eight weeks later. Spray with fungicide each month and occasionally allow the pot to dry out for a day, and by mid-summer flowers will appear. Two years after planting, plants will be about 40 cm (16 in) high. Small plantlets will often appear in midair amongst the branches of the flower stalks and these can be removed and planted up once they have achieved ten to twelve leaves. To do this, remove three quarters of the lower leaves, dip the stem in a rooting hormone and plant in a soil medium of two parts peat, one of sand. Roots should appear in one to two months. Use cell pots for seedlings to reduce root disturbance, a major killer of *Drosophyllum*.

Heliamphora

Heliamphora exists naturally with a climate of tropical humid days and temperate humid nights. Temperature requirements are, therefore, 18–23°C (65–75°F) day temperature, dropping to 2–10°C (35–50°F) overnight. Extended periods of temperatures in excess of 29°C (85°F) stresses the plants. To achieve ideal conditions, many growers in subtropic and temperate regions grow their plants in a glasshouse designed for *Nepenthes* highland species.

Growers in subtropical regions frequently use a soil mix of four parts peat and one of perlite or just sphagnum, or equal parts of peat, sand and leaf mould, and they provide high humidity by surrounding the pot with sphagnum moss or by using a fogger. Constantly using a fogger ensures the traps are full of water, aiding decomposition of prey. For growers living in tropical areas the greatest problem is keeping the plants cool during the day and night, best achieved using a fogger that activates at temperatures above 25°C (77°F). Growers in temperate regions, on the other hand, need to warm the air and provide humidity, best achieved by using an aquarium and aquarium water heater set above 5 cm (2 in) or so of water, on an open grate or tray on which are placed the *Heliamphora* pots. Ensure the pots are not in contact with the water. Light provided by fluorescent tubes above the aquarium, set on a cycle of 16 hours on, 8 hours off, ensures rapid growth. Such elaborate care is not necessary with plants four years or older, which can be grown successfully in a lowland *Nepenthes* glasshouse.

NEPENTHES

Lowland Species

If you live in the tropics or subtropics, and the humidity is high, *Nepenthes* can be grown outside. Depending on the species, full sunlight will produce large, full coloured pitchers. Bright sunlight may produce small red dots or blotches on the leaves, but do not be alarmed by this as plants in the wild also have these blotches. The same plants, in a filtered light position, produce greener leaves with small, less colourful pitchers. In cooler climates a glasshouse or terrarium is necessary.

Provided the days are warm (20–25°C (68–77°F)) many *Nepenthes* can withstand overnight temperatures as low as 10°C (68°F) for short periods of time, especially if plants are kept drier than usual. Under 'ideal' glasshouse conditions, plants tend to be 'softer', grow taller, produce less colour on the pitchers, have smaller pitchers and be more prone to insect and fungal attack than the same plant grown under harsher conditions. In glasshouses, pitchers are often not produced if fertilisers are used—it is as if the plant's system is aware that it is receiving ample food. A 'harder' environment, in contrast, produces thick, strong closely-packed leaves with sparse, very rigid, pitchers. *Nepenthes* do not like frosts so use a frost detector connected to an overhead sprinkler for plants grown outside; plants grown this way also require a drier soil during winter. In the tropics, an open soil such as crushed tiles or bark chips is best. In subtropical or temperate areas the usual growing medium is sphagnum in a hanging basket or equal parts peat and scoria in pots.

Highland Species

Highland species experience the cool nights of high altitudes, where the temperature can drop lower than 8°C (46°F) and seldom rises above 27°C (80°F). Consequently these species are hardy, and can be grown outside in tropical and subtropical climates. However, when grown in tropical climates with night temperatures that remain high, the plants tend to be less robust and produce softer leaves and less colourful pitchers than their natural counterparts. In temperate climates highland *Nepenthes* can be grown in a cool glasshouse along with *Dionaea* and *Sarracenia*. Plants grown in well-drained hanging baskets suspended close to the roof of a glasshouse grow better than those hanging lower down, though shade cloth is needed to prevent burning. *Nepenthes* can also be grown in large glass terrariums, which should contain about 8 cm (3.4 in) of water at the base and an aquarium or water heater. Suspend the plants above the water and set fluorescent tubes on a timer for 16 hours light each day, not more than 15 cm (59 in) above the plants.

Both highland and lowland species should be cut back every few years to about 30 cm (12 in). The cuttings could be used to propagate new plants or to graft onto healthier plants or less rare stock—the resulting plants will be stronger and produce lower, more attractive pitchers. The soil mix for highland *Nepenthes* is the same as for lowland species, except where crushed tiles and bark are used add one part peat moss.

Nepenthes x hookeriana

CULTIVATION

Pinguicula longifolia

PINGUICULA

Tropical Species

P. agnata is a typical example of a tropical species. In its natural outdoor bog environment it receives filtered light, growing amongst tall grasses and *Sarracenia*. This plant prefers high humidity and, given its eventual size, it is best potted into a 20 cm (8 in) squat pot. As for light requirements the light level on an open bench in a glasshouse is usually too much, while under the bench it will not receive enough. *Pinguicula* can grow very well, however, under benches that have fluorescent lights suspended more than 10 cm (4 in) above the plant in a 12 hour cycle. Alternately, cover the glasshouse with 20 per cent shade cloth.

Use a soil mix of two parts peat to one each of sand and perlite, with 2.5 cm (1 in) of sphagnum on top (this helps increase humidity and avoids splashing the flat ground-hugging leaves with dirt when watering).

Temperate Species

P. vulgaris is a widely distributed temperate species, growing in acid, neutral or alkaline soil. Plant it in a soil mix of one part peat and one of sand, sphagnum moss or one part peat and one of vermiculite. Place small pots in a tray of water with high humidity, full sunlight and warm day temperatures (20–30°C (68–86°F)). When sunlight, temperature and humidity are low a bigger pot with ample drainage (not a tray) is recommended. Plants with hibernacula should be kept drier than usual during winter and should be sprayed with leaf fungicide each month. Hibernacula can also be refrigerated during winter, then sprayed with a leaf fungicide before returning to the glasshouse sometime in spring.

SARRACENIA

All *Sarracenia* require full sunlight, a three to five month dormancy period, and ample pure water to produce adequately formed pitchers. If the temperature exceeds 35°C (95°F), humidity needs to be increased or the plants need to be sitting in trays of water. Most *Sarracenia* can be grown in bog gardens, bonsai dishes or water wells; they can also be grown in terrariums, and in glasshouses where there is no winter heat and the humidity does not exceed 70 per cent. Planted alongside *Cephalotus*, *Darlingtonia*, some terrestrial *Utricularia*, and *Drosera*, *Sarracenia* is ideal beside a dam or pond, or on a mixed-bench arrangement. On the mixed-bench, cover the surface with strong plastic, place sphagnum moss onto it, and add the plants directly to the moss. Seeds will fall to the moss and propagate, although the problem is that *Drosera* and *Utricularia* do tend to take over within a couple of months.

Healthy plants respond to nitrogen-based fertilisers fed through the watering system. When using tray water, feeding nitrogen direct to the roots via the tray is best. Ideally, remove all the pitchers at the beginning of winter (except for *S. purpurea* and *S. psittacina*) leaving only 2–3 cm (0.75–1.2 in) remaining. Towards the end of winter divide and repot into new soil which could be a mixture of peat and sand or just peat moss. Species such as *S. purpurea* and *S. psittacina* prefer soil of just sphagnum and a pot that is squat and gives the plant room to spread out.

Sarracenia minor: *note how wet the soil needs to be.*

Drosera paleacea

Propagation

Utricularia multifida

There are a variety of propagation methods available, and you should choose the appropriate one for each species and plant according to considerations of age, size, variety, climatic conditions and equipment available. Further, methods such as tissue culturing and stem cuttings—appropriate for a commercial nursery—may be inappropriate for the individual grower, who is likely to have a higher success rate with layering, leaf and rhizome cutting and grafting. As carnivorous plants are so varied, however, these methods vary according to requirements of each genus and species. For this reason, this section is divided into methods of propagation as demonstrated with *particular* species. For further information, consult the genus and species descriptions, as well as the Cultivation chapter.

Cuttings

Cephalotus

Leaf and pitcher cuttings provide a flowering plant in about half the time it takes to grow one from seed, and it is quite easy. Simply remove the leaf or pitcher as close to the rhizome as possible, dip the severed end into a rooting hormone and insert it about half way along the stem into a mix of three parts peat and one part scoria. Place the cutting in filtered light in a well ventilated area. As with seeds, spray with fungicides regularly, and repot in 12 months.

In both cases watering by tray results in plants rotting, so water from overhead or with high humidity in a closed chamber, in a filtered light position.

Another method of propagation commonly used by commercial nurseries and hobbyists is rhizome cuttings. This is best achieved by beginning with cuttings of healthy, thick, dark burgundy rhizome about 5–10 cm (2–4 in) long. Dip them in rooting powder and place horizontally about 2.5 cm (1 in) below the soil surface in a propagating tray. Water well, and drench well with a soil fungicide each month. As explained previously, use of a sterile soil is a good precautions (see p. 31). Provided cuttings can be kept cool, the best time to take leaf, pitcher or rhizome cuttings is the beginning of summer.

Darlingtonia

Seeds collected in late autumn should be kept quite dry and refrigerated until the beginning of spring, then planted by sprinkling on the surface of damp sphagnum moss. As with most carnivorous plant seeds a warm soil temperature will increase the germination rate and the rate of initial growth. Unlike many other carnivorous plants the soil temperature should be reduced once about four leaves appear. When planting seeds you have not collected yourself, determine when they were collected and whether they have experienced dormancy. At worst it is better to stratify longer than not at all. If your plants naturally experience temperatures as low as 5°C (41°F) for three months, then simply place the seeds around the soil of your *D. californica*, and repot as summer approaches.

It appears, even with tissue culture plants, that *Darlingtonia* needs to experience at least two winters for the non-juvenile pitchers to appear each year. Where I have seen adult pitchers appear 12 months after release from culture, the plant often turns back to juvenile pitchers in the second year, for three months or more then, towards the end of summer adult pitchers appear again. Pitchers

Darlingtonia californica: *dividing up the stolon.*

D. californica: *two months after division; note white shoot.*

D. californica: *three months after division; white shoot has developed into a plantlet.*

radiate from a central point, where there are two or more growth points. Divide along the rhizome separating these growth points, as each will have its own root system. Leaf cuttings, obviously, have no root system and so need to be grown where humidity is maintained, using either a terrarium, overhead misting system or (more primitive, and less effective) in a foam box sitting on a lower greenhouse shelf on a base of sphagnum moss. Make cuttings as close to the base of the rhizome as possible, ensuring that the white fleshy part of the plant is attached. Dip it into a rooting hormone and insert into the soil. The best time to take these and stolon cuttings is mid-spring: spray each cutting with a fungicide. As an added safety precaution water through a soil fungicide every four weeks for three months.

Electron microscope shot of the seed of a Dionaea muscipula.

Dionaea muscipula: *trimmed and ready to survive dormancy.*

Dionaea muscipula

As spring approaches and new traps begin to develop, leaf cuttings can be made by removing healthy mature leaves at the base of the rhizome (using a scalpel or very sharp pointed knife) ensuring that some of the white rhizome is attached. Dip this white section in a rooting hormone before placing the cutting in an environment with constant high humidity and warmth.

As cuttings tend to rot, a more open fresh soil mix than that used for adult plants is preferable, with three parts peat and two of sand (or three parts peat and two of vermiculite) and plenty of drainage. Do not let these cuttings sit in a tray of water. An ideal environment is a terrarium away from full sunlight. Within two months a plantlet, looking like a small *Dionaea*, will emerge at the base of the leaf. By now the old leaf will have died off and in a month of two the new plant should be potted up.

Non Tuberous *Drosera*

Propagation is possible for most of these species via leaf-cuttings, root-cuttings, tissue culture, seeds, and decapitation. If using leaf cuttings, remove the leaf as close to the base as possible, dip into a rooting hormone, and plant in a soil mix of one part peat and one part sand, and cover with a glass jar or place in a terrarium in indirect light. New leaves should appear within four weeks. Root layering is also a simple way to produce many plants, and is easily achieved when repotting. Instead of planting roots to point downwards, lay them horizontally to the surface—within two weeks new shoots will appear every 1–2.5 cm (0.4–1 in) along the root. Once four or more leaves have developed, the plantlets can be potted up and treated as normal. From seed to flower takes less than two months with root layering, three months with leaf cuttings.

Heliamphora

Leaf cuttings that are cut cleanly and as close to the rhizome as possible will root within four weeks and within two years produce adult-size plants, flowering in three years. From seed to flowering takes at least four years. Division is the easiest way of propagating, followed by tissue culture (which is very productive but quite problematic) and leaf cuttings. Propagation from seed is the least productive.

Nepenthes

If you wish to promote side shoots rather than height for your plants, remove the growing tip and apply a shoot-promoting hormone to slight incisions in the stem at each leaf. Buds form (depending on the species and conditions) within 3–4 weeks.

When using cuttings to propagate, they will produce a split and begin to bulge within two weeks and produce roots within four weeks. A good strike rate is achieved when cuttings are placed in sphagnum moss—strike in pure live sphagnum then repot after two months into the usual peat/sand mix. If you live in the tropics, however, with the associated high humidity and rainfall then cuttings can be placed in crushed bricks and chunks of bark. Alternatively, leave cuttings in a clear plastic bag half-filled with sphagnum, tied at the top and suspended among benches in a tropical glasshouse, or, if you do not live in the tropics, in a glasshouse with heated benches and misting systems.

Sarracenia

Sarracenia all grow at a similar rate. *S. leucophylla*, for example, attains a height of about 5 cm (2 in) in its first year, and 10 cm (4 in) the following year, about 20 cm (8 in) in the third year, and so on, increasing in height by about half the length of the previous year's height. From seed to flowering, the plant takes five years under good conditions. Once a plant is mature, there are many ways to propagate it. One of the easiest and quickest ways is

Sarracenia flava: *there are many ways to propagate* Sarracenia, *and it is a good idea to divide your plants at least every second year.*

to divide the rhizome with a sharp knife at its two growth points (two areas where new pitchers are simultaneously being formed). Another method is to remove the new growth (front section) of the plant so that two pieces exist, one with a few roots and all the pitchers, the other with just a rhizome and most of the other roots. The new growth continues to grow and send down new roots, and the back half with no pitchers will usually produce small clumps either side of the rhizome. Within a month these 'clumps' can be removed before they send down roots, dipped into a rooting hormone and planted about 1–2.5 cm (0.4–1 in) deep in peat moss and kept constantly wet and shaded. The rhizome will continue to produce the clumps in semi-shade, and they can continue to be cropped off the plant. Over 20 young plants can be obtained, which will flower in about 2–3 years, and respond quite well to bottom heat and misting. Another less certain method is to cut up the rhizome into 2.5 cm (1 in) sections, or to cut notches along the rhizome every 2.5 cm (1 in) and place these on the soil surface, lightly covering the rhizome with peat.

Whichever of the methods you use it is a good idea to divide at least every second year, if only to prevent the chance of losing your one and only specimen. *S. purpurea* and *S. psittacina* propagate easily from leaf cuttings, provided a small section of the rhizome is attached. From this method you can harvest one leaf every month, resulting in about six new plants each year.

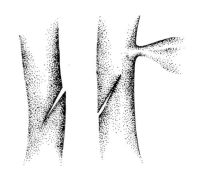

Grafting Nepenthes: *make a cut in both the stock plant (right) and the graft plant (left).*

Insert the tip of the stock plant into cut on the graft plant and bind together with grafting tape.

After about 30 days new leaves appear on the grafted plants. Remove the grafting tape.

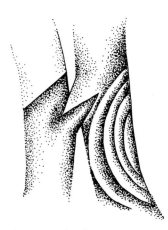

Sever the connection between upper and lower portions of the graft gradually. Chip away at the stem every ten days for about 40 days.

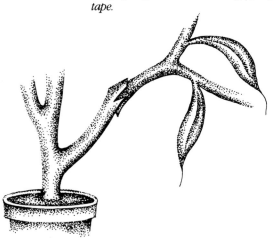

A final cut separates the graft from the original base: you have now grafted a new Nepenthes.

Grafting

Nepenthes

A useful way of propagating *Nepenthes* is via grafting. There are a number of reasons for attempting grafting, not least of which is to get to know your carnivorous plants better, and to attempt another method of propagation. Another and very important reason for attempting grafting is that it allows you to increase numbers of a prize plant, while safeguarding against its loss.

To attempt this method of propagation, you use what is referred to as a 'stock' plant, to which one or more 'graft' plants are grafted. First, trim back the stock plant so that only three stems, each about 15–30 cm (6–12 in) long, remain. Make a cut halfway through one side of the stem of all plants, and insert the tip of the stock plant into the cut on the graft plant. Now, bind them together with grafting tape.

After about 30 days, new leaves appear on the grafted plants. This is a sign to remove the grafting tape and to start severing the connection between the upper and lower sections of the graft. This is a delicate operation, and should be attempted in stages, gradually chipping away at the stem every ten days. After about 40 days, a final cut separates the graft from the original base: you have now grafted a new *Nepenthes*.

Multiple Division

Dionaea muscipula

Having already described the use of leaf cuttings to propagate this species, another method of propagation requires the newest section of an old rhizome (consisting of 2–3 leaves) to be cut away from the old rhizome every two months during the 6–8 month growing season. This can result in mature plants within one year, depending on the age of the parent bulb. Alternatively take a new strong leaf and make four or five nicks along the leaf, pin this notched leaf down on damp peat and wait for four or five plantlets to appear along the leaf.

AERIAL LAYERING

Aerial layering is a method useful for *Nepenthes*: to propagate the plant; because it is beginning to outgrow a terrarium or glasshouse; to encourage larger lower pitchers to develop, or to safeguard against loss or damage of your only species. Follow diagrams and instructions on this page, and note the photographs on following pages.

Entire Nepenthes *plant prior to aerial layering.*

Remove excess leaves and pitchers to strengthen the plant; make nick in stem as indicated.

Tie a plastic bag full of sphagnum moss over the nick, to encourage roots to develop. This will remain on the plant for two months, in which time the plant continues to grow.

Divide the plant just below bag of moss once roots are established. The bottom half of the plant, illustrated here, will continue to grow.

Remove plastic bag from solid mat of roots and moss, and repot top half of plant, which will continue to grow.

Nepenthes maxima *before aerial layering. Various propagation methods are available for* Nepenthes, *all of which also need to deal with the extreme height of this genus.*

Nepenthes maxima *laid out and divided in an attempt to make more manageable plants. Cutting is a useful method for propagating* Nepenthes.

Nepenthes maxima *cutting showing roots and shoots, two months later.*

TISSUE CULTURE

The various genera of carnivorous plants react to tissue culturing individually, as might be expected. Multiplication rates vary, from the very prolific *Drosera* to the least prolific *Darlingtonia*. *Dionaea*, for example, multiplies at a rate of eight times every four weeks, so that in theory over one million plants could be propagated from one tip or plantlet in one year.

This method of plant propagation is best achieved in a sterile environment, although some hobbyists have successfully cloned or cultured at home, without sophisticated equipment. Often, too, the preparation of tissue culture actually takes place in laboratories, and you can then follow through by planting out the plants raised by this method yourself. This is a rewarding method, however, and I have outlined the general procedure.

Tissue culture is appropriate for both seeds and existing plant tissues, and both have their benefits. Seeds tend to be easier to work with, but they are much more variable; culturing from living tissue is difficult, but you know what your end plant will be. It is a good idea to begin with seeds, and progress to living tissue. When using living tissue, it is important to first choose plants that are healthy and free of disease. The method outlined here is appropriate for both seeds and shoots:

1. Ensure your work area, and hands, are thoroughly clean.
2. Working quickly in a stable environment, prepare your multiplication medium according to instructions and pour into a suitable container. The first media involved is called a multiplication media and consists of nutrients and hormones in agar, which work to promote shoot development and inhibit root development.

Plants propagated via tissue culture in a convenient container. They grow on a mix of agar and multiplication medium.

Plants propagated via tissue culture (l to r): Byblis gigantea, Cephalotus follicularis, Dionaea muscipula, Drosera spathulata *and* Sarracenia flava.

Hundreds of Dionaea muscipula, *propagated via tissue culture, multiplied into tubes and awaiting planting out.*

Pinguicula caerulea *after being planted out from tissue culture.*

3. Once the solution has cooled, sprinkle on the seeds or pollen grains, or place meristem (prepared according to step 4) onto it.
4. If working with living tissue, remove about 2 cm (0.75 in) of a growing tip from any part of your carnivorous plant (newest shoots are best). Next, soak the shoot in a bleach solution for 30 minutes, then rinse in sterile water. In a laminar flow cabinet, slice off 2 mm (0.08 in) from the tip of your shoot—this is called a meristem. The meristem is now ready to be placed on the agar.
5. Position the containers or test tubes no more than 23 cm (9 in) from artificial light in a temperature of 20–24°C (68–75°F). After three to six weeks a mass of shoots will develop.
6. Divide the plant clumps in a lamina flow cabinet after a further three to six weeks. Place the plantlets into planting out containers about 12 cm (5 in) wide and 6 cm (2.5 in) deep, filled with 2 cm (0.75 in) of rooting or multiplication medium. Rooting medium is one that suppresses shoots and promotes root growth. It is necessary to transfer plantlets to the rooting medium at this stage as plants grown exclusively in multiplication medium grow slowly and are weak. Different media are appropriate for different stages of growth.
7. After another three to six weeks plantlets are ready to be planted out. When *Dionaea* propagated in this way are 16 to 20 weeks old, for example, they have the appearance of a three-year-old plant grown from seed.

Planting Out (Weaning)

As plants propagated via tissue culture exist in a rarefied environment, when transferred they are vulnerable, and may dry out, wilt or be attacked by fungus. For this reason, fungicides are very helpful. To protect against this, plants must be 'hardened off' gradually, by using a systemic fungicide—many fungicides have both curative and preventative action, the latter achieved by residual activity. Halting the spread of fungal activity requires strong and frequent doses of fungicide, and removal of all infected material.

To plant out from tissue culture, use the following method:

1. Carefully remove all plants from container and place in another container of lukewarm water that contains a weak solution of fungicide.
2. Ensure all agar is removed from the plants by agitating them in the water, then place them in sterile soil. *Do not use old or pre-used soil.*
3. Spray with fungicide every week for four weeks.
4. Place the tray in a drought-free high-humidity environment, such as a terrarium. The plants should be shaded by 50 to 90 per cent shadecloth, with a humidity level of 70 to 100 per cent and within temperature ranges of 19–25°C (66–77°F). After about two weeks the plants can cope with temperature and humidity fluctuations and the lid of the terrarium can be lifted slightly, and relative humidity of 60 to 80 per cent is adequate, at temperatures of 15–25°C (59–77°F). If plants are looking limp, prolong the period of hardening off by two or more weeks, or until the plants pick up again.
5. Maintain this environment for two weeks, after which the tray can be placed in a glasshouse under normal conditions. Temperatures exceeding 30°C (86°F) should be avoided, as should drying out of the plants and soil.
6. After six to eight weeks the plants will be able to cope with full sun and can be grown according to their regular species requirements.

SETTING UP A GLASSHOUSE

Both the design and materials appropriate for each glasshouse depends on the space available, the position of the glasshouse and the funds available. Guiding principals, however, can be ascertained by getting together a list of the plants you intend to house, and noting their climatic, environmental and other, requirements.

During the day when the sun is shining directly on the glasshouse, temperatures can rise dramatically, as much as 5 or 10°C (41 or 50°F) above the outside temperature—thus, a 'greenhouse effect'. As many carnivorous plants cannot endure temperatures above 35°C (95°F), something needs to be done to control the temperature. Options include extractor fans, ridge vents and shade cloths.

Difficulties arise when you leave the house before 9 am and return after 5 pm, by which time the temperature has returned to a reasonable level. I have often been greeted by a rush of hot air when I open the doors late in the morning or early afternoon. Night temperatures on the other hand can vary by as little as 2°C (35°F) from the outside temperature, which for plants such as *Sarracenia*, *Dionaea*, *Byblis*, some *Drosera*, *Utricularia* and *Cephalotus*, is quite acceptable. For some of the more tropical species, however, this can be a problem, and some form of artificial heating is usually necessary.

To maintain a temperature of 15°C (59°F), you might use an electric fan heater, hot water heater or soil heating, etc. Hardy plants such as *Drosera*, *Sarracenia*, *Dionaea* etc have seeds that will germinate at 15°C (59°F), while tropical plants such as *Nepenthes* and some *Utricularia* require 21°C (69°F) to germinate. If your tropical collection is small it may be sufficient to heat only a small part of the glasshouse: use plastic sheeting as a divider, and a small area heater (such as a piglet heater or light globes), or else a glass tank with its own heater could be set up quite easily.

To obtain an even spread of light, position the glasshouse in a north–south direction, and keep the covering material free of dust, as dust reduces light intensity and causes irregular lighting.

When maintaining a fairly warm temperature is vital, use the best covering material you can afford—I prefer those with an air layer between the two sheets. Ensure that all spaces and cracks are sealed, as small cracks can reduce the temperature by as much as 5–10°C (41–50°F).

The range of materials used for flooring is extensive, and depends largely on personal preference: scoria (perlite), pine bark chips, sawdust, concrete or bricks are all suitable. Essentially, the material should be able to absorb a certain amount of water that can be released later to provide humidity when the temperature increases. Thus, on hot days it is the floor that needs to be well moistened to prevent the air from becoming too dry. It is best to line the floor with heavy duty plastic before applying any flooring material.

Assorted Sarracenia *growing in a glasshouse.*

Humidity and high temperatures are important then, but while a lack of draughts keeps the temperature up, good ventilation is required to prevent the germination of fungus spores. It also increases carbon dioxide, which aids in photosynthesis (some nurseries even go so far as adding carbon dioxide artificially to promote growth). Ideally, you should strive for some type of vent at floor level to suck in cold air, and a roof vent to expel hot air. Having two automated vents working 24 hours is a luxury few growers can indulge: an option, therefore, is to (inexpensively) open louvres or doors.

In areas of high natural temperatures the grower's problem is not maintaining heat, but rather finding ways of reducing it. Consequently, tropical species can be grown outside under shadecloth. Or, on days of extreme heat in a glasshouse receiving full sun, it may be necessary to have shadecloth over the whole house. If this is necessary use 50 per cent mesh positioned about 22 cm (9 in) away from the glasshouse on all sides including the top. It may be necessary to build a separate frame to ensure this gap is constant.

Watering methods in your glasshouse will depend on the plant species involved, size of pots, the soil mix, amount of humidity, and amount of sunlight. Take all of these factors into consideration when you choose your watering methods. Even with the best automatic system, it will be necessary to individually water some plants. This can be lessened by keeping the same species and same size plants and same size pots in one area.

Plants from tropical areas may need to be watered by a misting or fogging system, automated by either an artificial leaf or some type of balance arm. While misting systems tend to work well, there can be long-term problems of accelerated growth rates, leading to weaker and more delicate plants, plants that are too tall and stringy, and undersized pitchers. A better system may be to use the misting system for only a few hours each day, and water each pot by hand early in the morning.

A glasshouse is not the only solution to growing carnivorous plants. Choices depend upon climate and space, and the range of species you intend to grow. Of course, there are also disadvantages involved in using glasshouses, not least of which is possible problems with mildew, aphids, scale and slugs.

INSECTICIDES

A whole range of insects attack carnivorous plants, including aphids, mealy bugs, red spider mites, thrips, whitefly, caterpillars, ants, snails, wasps, slugs and *Sarracenia* root borer. Glasshouses with their ideal conditions and vulnerable plants make a good home for these pests.

Changing conditions to prevent attack, rather than using insecticides, is the ideal approach. Thrip and red spider mite, for example, can be eliminated by dramatically increasing humidity, while the alternative is to use Metasystox and Pyrethrin. Another means of changing conditions to combat pest problems is to lower plants into a trough of water, covering all foliage parts, and leaving it for two hours. As a simple solution, snails and slugs can usually be spotted and removed late at night. It is also worth

Another 'sick' Sarracenia, *showing the effects of poor environment, pests and variations in light and water supply.*

exploring environmentally-friendly alternatives, from prevention to the use of organic substances like Pyrethrin.

If you really do want to control a severe outbreak by using insecticides, then there are some consequences. Using White Oil and Malathion to remove aphids and mealy bugs, for example, can cause a distortion of leaves and pitchers.

FUNGICIDES

A practical way of preventing both insect and fungal attack is to ensure the growing area is kept free of all dead or broken leaves, that only new pots are used and that any affected plant is immediately removed from the collection. For commercial nurseries with thousands of plants, more extreme prevention is often preferred. This is carried out by spraying foliage with a leaf fungicide every month and a soil fungicide every two months, with a follow-up dose applied one week later.

If fungus attacks plants, the leaves start to develop white spots and quickly turn into larger black or brown areas that eventually cover the whole plant. Badly affected leaves should be removed and burnt, or placed outside during the hot dry summer months.

Care should always be taken when using fungicides, as they can be extremely dangerous.

A Sarracenia *showing the effects of variations in light and water supply.*

Carnivorous Plant Genera

Drosera oreopodion

Aldrovanda
ăl-dro-'văn-dă

Aldrovanda, of the family Droseraceae, is a monotypic genus with four known varieties. It is a rootless plant that floats in shallow fresh water, just below the surface, in amongst reeds and rushes, in swamps, lakes and dams. It is native to east and central Europe, Africa, Japan, northern Australia and India and was named by Monti, an Italian botanist, in 1747 after Ulisse Aldrovandi, the Italian botanist responsible for setting up the Botanic Gardens in Bologna. Charles Darwin called *Aldrovanda* a miniature aquatic *Dionaea* and it too has sensitive trigger hairs either side of the traps. While there are usually three trigger hairs on each side of *Dionaea Aldrovanda* has about 20 each side.

Aldrovanda is commonly called the waterwheel plant because it is like a series of wheels, the spokes (leaves) having small traps on their tips. Each wheel has eight spokes, each terminating in a trap, all of which face one direction. Any one plant has 100 or more traps. It looks like a light green, semi-translucent, hairy caterpillar 15–25 cm (6–10 in) long and 2 cm (0.8 in) wide. The traps are similar to *Dionaea* 6 × 4 mm (0.25 × 0.16 in), with four to eight bristles, each as long as the traps. It preys upon water fleas, daphnia and other small micro-organisms. The trap closes in 0.02 seconds but complete tightening is slower, taking many hours and requiring continual stimulation of the trigger hairs. The trap opens and is reset after several days, unless the prey is very large, in which case the trap may remain forever closed.

Aldrovanda vesiculosa

The system causing trap movement with inner and outer walls tension similar to *Dionaea*

In cool climates the rear of the plant dies away while the front continues to grow. In warmer climates the plant branches out, producing offshoots.

During winter, where the water temperature falls below 17°C (63°F), *Androvanda* will form winter resting buds, sink to the floor and float up again as the temperature increases. Under ideal conditions (water temperature 25°C (77°F)), plants produce a single small white flower with five petals. It appears on a short scape during spring in warmer climates and in summer in cooler climates. It needs to be artificially fertilised, and if this succeeds the flower will eventually produce six to 20 oval seeds.

Aldrovanda grows well with moderate care in a water container such as an oxygenated water fish tank. Australian *Aldrovanda* are larger than those from Europe, and those from India are larger still.

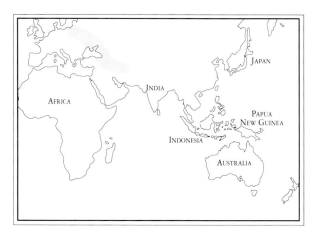

Aldrovanda vesiculosa

Aldrovanda vesiculosa

Common Name: Waterwheel Plant
Size: 15–25 cm (6–10 in) long.
Climate: Tropical.
Habitat: Freshwater swamps and dams.
Distribution: Africa, northern Australia, east and central Europe, India and Japan.
Flowering Time: Spring to summer.
Colour: Traps and leaves green; flower white.
Description: *A. vesiculosa* is an aquatie species with over 100 traps. The small bristly traps appear at the end of spoke-like leaves that grow in a wheel shape. The single white flower has five petals and appears at the end of a short scape, and produces 6–20 seeds; in warmer climates it reproduces via offshoots. In dormancy this species produces winter resting buds.

Brocchinia
brō-kĭn-ē-à

Brocchinia has five species, named by J. Schultes in 1830 in honour of Giovanni Brocchi, director of the Botanic Gardens in Brescia, Italy. *Brocchinia* is one of a number of genera within the Bromeliad family. To date only *B. reducta* and *B. hectioides* have proved to be carnivorous. It is a recent addition to the list of carnivorous plants and was added in 1984. *Brocchinia* is remarkable, as it is the only genus to include both carnivorous and non-carnivorous plants. Even more remarkable, within the leaves of *B. reducta* exists another carnivorous plant, namely *Utricularia humboldtii*.

B. reducta and *B. hectioides* are native to Venezuela and exist in the lowland savannahs surrounding tepui as well as on the tepui themselves. Like many other Bromeliades, all *Brocchinia* have a woody trunk, with leaves encompassing a central well of water. It has a pitfall trap, with waxed walls and a lower digestive zone. The leaves are 50 cm (20 in) long and 5 cm (2 in) wide: bright yellow with curled up edges in full sunlight, they are green with non-curled margins when they grow in less light. The many small flowers are 5 mm (0.16 in) long, white and solitary on an inflorescence that can be up to 60 cm (2 ft) high. The inflorescence is a many-branched panicle, 20–30 cm (8–12 in) long. The seeds are numerous and flat. *Brocchinia* grow naturally in clumps and are spread over many areas in a wet soil mix of five parts peat and one of sand, within a temperature range of 18–35°C (-7°F–1°F) under high humidity.

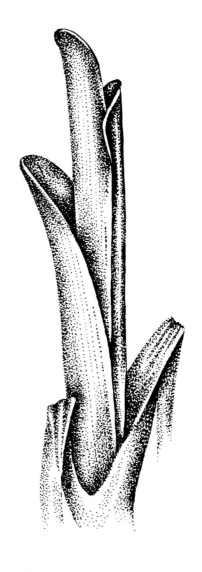

Brocchinia reducta

Brocchinia reducta

Brocchinia reducta

Common Name: None.
Size: Up to 60 cm (2 ft) tall.
Climate: Humid.
Habitat: Lowland savannah as well as on the tepuis.
Distribution: Venezuela.
Flowering Time: At any time in the wild, or in spring or summer under cultivation.
Colour: Leaves green, or yellow in full sunlight; flowers white.
Description: *B. reducta* has a woody trunk, and leaves that curl up in full sunlight. It has a pitfall trap, and many flowers on a long inflorescence. It grows in clumps, in areas of high humidity.

Brocchinia reducta

Byblis
'bib-liss

Byblis contains two species—*B. gigantea* and *B. liniflora*—both native to Australia (*B. gigantea* is found only in Western Australia, and *B. liniflora* is common to the three northern states of Australia, and New Guinea).

Part of the Biblidaceae family, this genus was named in 1808 by R.A. Salisbury after Byblys, the Greek nymph who fell in love with her brother. Byblys' love was unrequited: she hanged herself and was turned into a fountain—it is the droplets of water that make this association relevant. It was another 31 years before John Lindley published the second species in this genus, *B. gigantea*.

Byblis have green thread-like leaves with clear sticky tentacles, tentacles that even appear on the flower stems. *Byblis* is often confused with *Drosera*, but the tentacles on *Byblis* are not dotted with red, the floral arrangement is different, and *Byblis* has triangular tube-like leaves in cross section.

The easiest species to grow is *B. liniflora* which, when treated like an annual, will grow from seed each year. It should be grown similarly to *Sarracenia* and *Dionaea B. gigantea*, on the other hand, is difficult to grow year after year. Older *B. gigantea* plants can, for no apparent reason, turn black and die.

Both species are very efficient at catching large moths and mosquitoes.

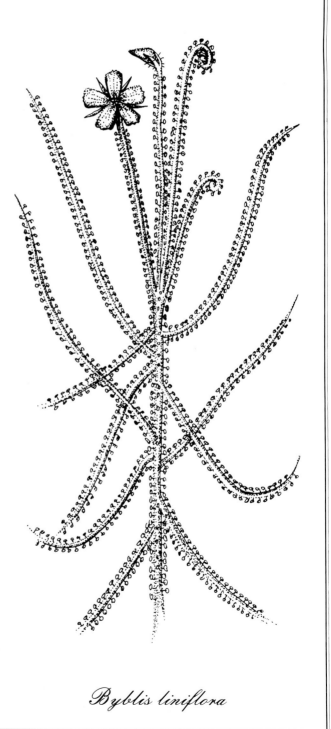

Byblis liniflora

Byblis gigantea

Common Name: Rainbow Plant.
Size: Up to 60 cm (24 in).
Climate: Subhumid, warm temperate.
Habitat: Sandy swamps.
Distribution: South west coast of Australia.
Flowering Time: Early spring in dry areas, early summer in wetter areas. In cultivation, usually flowers in mid-spring.
Colour: Woody brown base; leaves light green at tip and darker at base, and can appear very pale due to droplets of moisture that cling to the leaves; leaves can appear black, being covered in trapped insects; flowers are purple-blue in the centre fading to lilac on the petal fringe.
Description: An erect species with a strong base, able to withstand strong winds. Its slender triangular leaves can grow up to 30 cm (12 in) in length, and the numerous flowers (10 or more) are 3 cm (1.2 in) across. The five petals are surmounted by five green sepals connected by a single slender green stem. The five stamens bend inwards above an ovary that produces 50 or more seeds. Unlike the flowers of the *Drosera*, that grow in the same area, those of *B. gigantea* open as the daytime heat builds and close as it cools.

In areas where the soil dries out in late summer *B. gigantea* is an herbaceous perennial. It gradually deteriorates, turns black, becomes fragile and can break and blow away from its base. A couple of months later (usually late autumn to early winter) new shoots, usually three or more, will emerge through the soil.

In damper areas and under cultivation, it is a woody perennial, maintaining its leaves. *B. gigantea* benefits from fire, as it increases germination and reduces competition.

Tentacles on the flower stem catch crawling insects, so pollination is only possible with the help of flying insects.

Byblis gigantea

Byblis liniflora

Common Name: Rainbow Plant.
Size: In its natural habitat, up to 15 cm (6 in); in cultivation, up to 30 cm (12 in).
Climate: Humid, tropical.
Habitat: Wet, sandy soils.
Distribution: Northern Australia and New Guinea.
Flowering Time: Spring to summer or, in cultivation, summer to autumn.
Colour: Pale yellow to green leaves; flowers pale lilac.
Description: Easily grown from seed, this is a smaller, stringier plant than *B. gigantea*, usually no taller than 15 cm (6 in). Flowers have the same shape as *B. gigantea* and are 1-2 cm (0.4-0.75 in) in diameter. This species is usually treated as an annual as taller, older plants, lasting more than one season, are usually less attractive and require support. Often used in orchid glasshouses as a natural pest control.

Byblis liniflora

Catopsis
kă-'tŏp-siss

Catopsis has 21 species and belongs to the Bromeliad family (as does *Brocchinia reducta*), and to the subfamily of Tillandsioideae. All species are epiphytes or lithophytes, but only one, *C. berteroniana*, has been found to be carnivorous. There are still some species to be described, and it may be that some of these are also carnivorous. *Catopsis* was originally described as *Tillandsia* and *C. berteroniana* was called *Tillandsia berteroniana* by J.H. Schultes in 1830. In 1864 Griseback reclassified it as *Catopsis*. *Catopsis* comes from the Greek *katas* meaning beneath and *opsis* meaning appearance, implying the plants are seen from below. Flowers are often single sexed, with male and female plants having a different appearance, making description difficult.

All *Catopsis* have a rhyzome with offshoots, called 'pups', clustering around the main plant. These pups usually develop before flowering and should be removed and replanted, encouraging more pups to develop. Each plant usually flowers once only, three years after seed is sown; pups on the other hand will flower within twelve months of removal. After flowering, the plant's leaves turn yellow, wither and fall off. The leaves are clustered to provide a watertight 'tank', which becomes an aquarium. The seeds have tufts of hair and can be sown within a temperature range of 22–25°C (72–77°F), and shoots will appear within three weeks. *Catopsis* can be grown in a similar way to *Nepenthes*, in open wooden slatted baskets filled with sphagnum moss (or, if humidity and temperature is high, with bark chips). Alternatively, make up a wire mesh tube filled with sphagnum and attach *Catopsis* to it. Plants grown in pots should have a damp soil mix of one part peat moss to one part sand, or one part peat, one part sand and one part pine needles. This species responds well to regular fertilising and to having water poured down the funnel. Unlike many other carnivorous plants, however, *Catopsis* will die if sprayed with white oil.

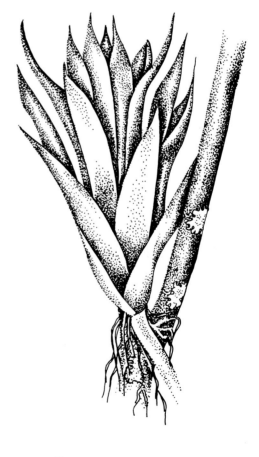

Catopsis berteroniana

Catopsis berteroniana

Catopsis berteroniana

Common Name: None.
Size: Up to 45–90 cm (18–36 in) high.
Climate: Tropical, humid.
Habitat: At altitudes of 3500 m (11500 ft), in full sunlight, high on the sides and forks of trees.
Distribution: South America, southern USA, West Indies.
Flowering Time: Spring to summer.
Colour: Leaves yellowish green; flowers white.
Description: *C. berteroniana* is an epiphyte that often grows on conifers. Unlike many Bromeliades, it has no spines on the edges of the leaves, but it does share characteristics of its family with its base-shape and waxy leaves. Leaves are 40 cm (16 in) long and 4–5 cm (1.5–2 in) wide, with an underside covered with white powder—leaves that are widest at the base also turn white at the base. The scented flowers are up to 2 cm (0.75 in) long and appear on a tall branching erect scape, up to 135 cm (53 in) long.

Cephalotus
sĕf-à-'lō-tŭs

Cephalotus, like *Dionaea*, is a monotypic genus, one of six in the carnivorous plant family. It is a member of the Cephalotaceae family, and the first published account of it was made by La Billardière in 1806. The name is derived from the Greek, *kephale*, head, and Latin *folliculus*, meaning small pod or bag. This refers to the head of pollen at the end of the stamen, and the pod-like pitcher shape.

C. follicularis is native to Western Australia, and is one of the easiest species to grow in tropic to temperate climates, especially once the plants are mature.

All stages of this species, even very small seedlings, produce pitchers, which have the appearance of small moccasins. In their juvenile stage the pitchers are covered in coarse white hairs, that protect the pitcher from harsh sunlight and drying out as it grows. As it grows larger, and as the connecting stem gets longer, the pitcher gradually fills to about one-third with water that tastes rather like commercial mineral water. Eventually the pitcher lid opens and is ready for prey.

As the pitcher grows to its mature length of 5 cm (2 in), the rim and lid become progressively darker shades of burgundy. Plants grown outside all year amass these pitchers during winter. The pitchers have a main central rib, and two side ribs that slope toward the opening. One theory is that these help guide insects toward the opening, however I have observed insects crawl up, across, and even down the ribs: I don't believe they provide a path but rather provide increased rigidity to the pitcher, especially to the exposed front. It is surprising how strong these thin pitchers really are. It takes considerable pressure to crush a pitcher, either around the circumference or from above, which has added to the plant's survival in an area grazed by kangaroos and, more recently, trampled by cattle.

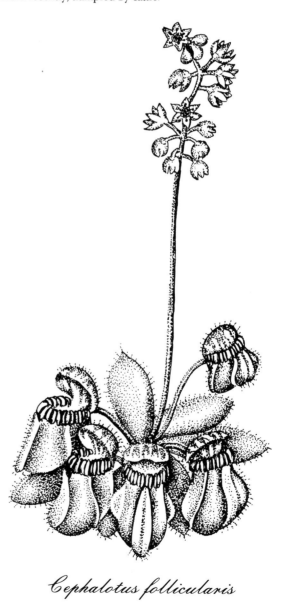

Cephalotus follicularis

Cephalotus follicularis

Common Name: Albany Pitcher Plant.
Size: Pitchers to 8 cm (3 in), scapes to 80 cm (32 in).
Climate: Subtropical, semi-humid.
Habitat: On the rise, or drier, side of peaty swamp land.
Distribution: Western Australia, on the south west coast of Australia, limited to a 250 km (155 ml) coastal strip from Cape Richie to Augusta.
Flowering Time: Summer. In cultivation flowering occurs earlier, so that by spring the 60 cm (24 in) long flower scape has withered and small brown pods of hairy seeds emerge.
Colour: Ovate leaves light to dark green in colour as they emerge, older leaves range from green to burgundy during mid-winter. With increasing sunlight the pitchers change in colour from green to dark red on the rim and the lid ribs, while increased sunlight makes the whole pitcher burgundy.
Description: *C. follicularis* is often found clustered amongst tall (60 cm (24 in)) grass tussocks, with roots intertwined. The tall grasses provide filtered light, which, in very dense areas, produces dark green pitchers. Where more light is available the pitchers are dark burgundy, strong, thick and waxy. During summer, the peaty sandy soil can be quite dry, while it can be flooded during winter. Winter can be cold and damp.

As summer approaches non-carnivorous leaves begin to form. They are thick (1-2 mm (0.04-0.08 in) and waxy and, like the pitchers, have white hairs, this time around the outer edge of the leaf and on the stem. As the leaves grow these fine hairs disappear. About six leaves and pitchers radiate from the centre of the plant, the pitcher's openings facing outward. As the plant produces more pitchers and leaves, a second cluster develops, and, beneath the soil, the main (dark brown to burgundy) rhizome produces a secondary (creamy tan) rhizome. The secondary rhizome can be removed when the pitchers have reached full size, using a sharp knife, where the secondary rhizome attaches to the main rhizome (by this time at least one single fine root will have developed on the secondary rhizome). The plant will continue to multiply by developing these clusters until pitchers and leaves become intertwined. Pots become so crowded that pitchers start touching the sides, or growing out of drainage holes in the bottom of pots. Because of this uncomplicated method of asexual propagation, in the wild *Cephalotus* tends to grow in scattered clusters.

For flowering to occur the rhizome needs to be about 7 cm (2.75 in) long and 0.5-1 cm (0.2-0.4 in) thick, with five to ten pitchers and 10 to 20 leaves. Flowers are small and inconspicuous, less than 1 cm (0.4 in) wide. The six waxy sepals are white at the tip, looking like small white petals, although the flowers in fact have no petals. Clustered inside the sepals are the twelve stamens, six tall and six short. As the flower stem emerges from

Cephalotus follicularis

Cephalotus follicularis

Cephalotus follicularis

Cephalotus follicularis

the centre of a cluster of leaves and pitchers it has the appearance of a small pitcher about to develop—it is not until this very hairy juvenile bud rises above the leaves and pitchers that you realize a flower is developing. As the scape grows taller, the apex thickens and, after a height of about 30 cm (12 in) a branch may develop and produce six or more flowers while the main group of flowers continues to grow. More branches develop until it is about 60-80 cm (24-30 in) tall, at which time (some 30 days from emergence) flowers continue to open. By now 30 to 50 flowers may have emerged, and three or more scapes. Younger plants may only produce a single scape with fewer than five flowers. It is advisable to remove any scape that may occur in the first year, to increase chances of survival, and for increased fertilisation cross-pollinate the flowers.

The seeds are 1 mm (0.04 in) long, encased in a 4 mm (0.16 in) long tan hairy husk, and is very light. Seeds ripen towards the end of summer and are dispersed by the wind, although this is not as effective as rhizome outcrops.

DARLINGTONIA
dar-ling-to-ni-ya

Darlingtonia is a member of the Sarraceniaceae family. John Torrey's detailed description of *Darlingtonia* was first published in 1853. He wanted to honour his friend and botanist Dr William Darlington by naming a plant for him. So desperate was Torrey to honour Darlington that he named another genus after his friend. As Darlington had at least three genera taking his name, and as the rules of International Botanical Nomenclature prohibits homonyms, the plant's name was changed to *Chrysamphora* in 1891. Finally, the other two genera that bore Darlington's name were found to be synonymous of older generic names, so *Darlingtonia* could be reinstated 63 years later.

Darlingtonia is an herbaceous perennial with tubular pitchers. It often grows to 1 m (3 ft) tall, emerging from a rhizome growing horizontally along the ground. As the pitchers emerge they face the centre of the plant, and while they grow the top twists 180 degrees, so that the opening with the fishtail shaped fangs points outward from the plant (removing the 'fishtails' results in the plant capturing less prey). Once the full 180 degree turn is complete the dome lifts up, the fangs spread out and the head puffs up, (hence the name 'Cobra Lily') revealing the entrance to its unusual spiral trap.

As the roots must be kept cool, it is difficult to grow *Darlingtonia* in subtropical climates, and virtually impossible in tropical ones. Temperate areas provide ideal conditions, making the species easy to grow, and flowers should appear every season.

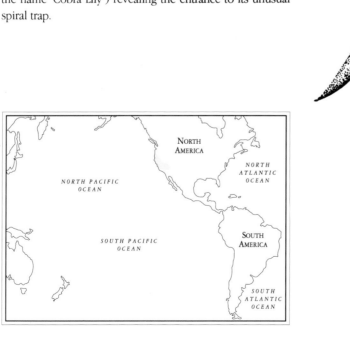

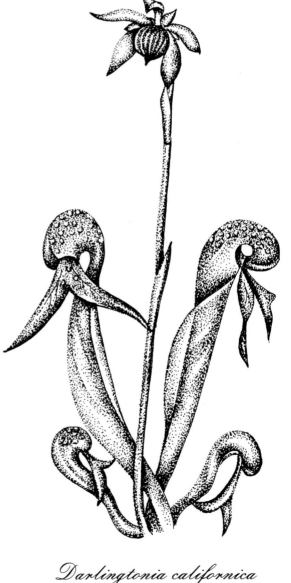

Darlingtonia californica

Darlingtonia californica

COMMON NAME: Cobra Lily, Californian Pitcher Plant.
SIZE: Up to 1 m (3.3 ft).
CLIMATE: Temperate to boreal, subhumid.
HABITAT: Both high and low altitudes. Around bogs, streams and springs and areas, at low altitudes in mountainous areas, whereever cool water is easily accessed by the shallow root system. The plants grow in sphagnum moss, and rock cracks in rapids and waterfalls at high altitudes (up to) 2800 m (9186 ft)).
DISTRIBUTION: Northern forests of the mid-west coast of North America from western Oregon to northern California.
FLOWERING TIME: Mid-summer at lower altitudes, to the end of summer at higher altitudes.
COLOUR: Pitchers light green at the base, changing to shades from green to dark burgundy on the upper third of the pitchers, depending on the light. As the colour changes up the pitcher fenestrations develop. Below the hood of the pitcher 'fangs' flair out and are usually the first to take on any colour. Where these plants are grown in full sunlight the pitcher can be totally burgundy by the end of autumn, and even with less light the pitcher tends to be blotched red and brown, unlike its response in the gentler and controlled environment of a glasshouse.
DESCRIPTION: When the pitchers develop from seed or young stolon (runner) only juvenile leaves develop. These leaves do not twist or have a hood, but are simply tubular with a primitive opening at the top, similar to a *Sarracenia* seedling. As the plant grows, usually after the first year, the juvenile leaves disappear and the typical *Darlingtonia* pitcher begins to develop. For the first three to five years the pitchers stay mostly prostrate, appearing to grow along the ground, with only the last third pointing upwards. Suddenly a new generation of pitchers will twist and point straight upward from the base of the pitcher thus having the effect of doubling the plant's overall height.

Both the entrance (5 cm (2 in) across) and the fangs point downwards, preventing rain from entering the pitcher. Plants grown outside produce thick, tough pitchers, reminiscent of the northern species of *Sarracenia purpurea*—the similarity is not surprising as both can be covered by snow during winter. As with *Sarracenia minor*, small translucent fenestrations on the hood give its interior the appearance of a well-lit glass dome. With so many windows of light it is no wonder insects become confused, buzz around in this mirror maze looking for the entrance, and eventually become trapped.

As *Darlingtonia* develops it produces stolons that can grow to a length of 30 cm (12 in) or more. They send up plantlets and roots while still attached to the parent plant, contributing to the dispersion of plants. Thin fragile roots emerge the full length of the rhizome. When these shallow roots do not receive water the plant can begin to wither within a day, resulting in pitchers that feel as soft as tissue paper. Provided the plants are watered during the night, the pitchers will be back to normal by early morning. When plants are kept in filtered light the pitchers are brittle and dark green, and more prone to insect and disease attack—increase the light and the leaves turn a more healthy yellow-to-burgundy colour.

Flowers of *D. californica* are described as 'noddling', which is to say they hang down, from a scape 60-100 cm (24-39 in) long that rises to 15 cm (6 in) or more above the pitchers. Three or four bracts (yellow through to burgundy) appear the length of the scape to encase the flowers as they develop. The flower has five petals with deep red to burgundy veining encased inside five yellow to pale green sepals. The petals all narrow abruptly about two thirds of the way along their length, so that two petals side by side make it appear that there is a hole in the petals. Once the sepals have moved to a horizontal position, and the petals begin to wither and drop off, the 15 stamens surrounding the top of the yellow and green ribbed bell-shaped ovary can be seen. An insect would be able to see the stigmas at the base of the bell through the small 'hole' of the petals. Entering, the insect brushes past the stigmas, leaving a pollen deposit from other flowers, on its way to the top of the bell. In order to further encourage cross-pollination the bell shape works like an umbrella, shielding the stigmas from fallen pollen. Some flower pods from previous seasons remain intact until mid-spring, when the bright sunlight makes the brittle pods disintegrate, releasing the club-shaped light brown seeds (pubescent at one end and thinner and hairless at the other).

A slime mite—*Sarranceniopus Darlingtonaea*—exists in the pitcher plant to eat any food trapped.

D. californica is often found alongside *Pinguicula vulgaris*, *P. macroceras* (both of which flower at the same time) and *Drosera rotundifolia*.

Darlingtonia californica

Dionaea
di-ˈō-ni-ȧ

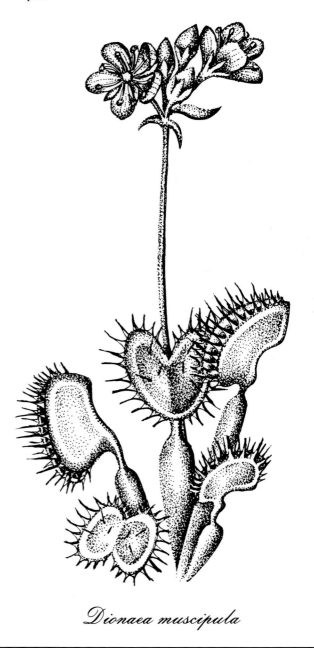

Dionaea is part of the Droseraceae family, and is a monotypic genus. *Dionaea*, named by John Ellis, is derived from ancient Greek mythology. Dione was the mother of Aphrodite, goddess of beauty and consort to Zeus: Ellis must have been truly impressed to give the plant this name. *Dionaea* gains its common name, Venus Fly Trap, from Dione's Roman counterpart, Venus. John Ellis first published a description of *Dionaea* in 1770.

Dionaea is one of the hardiest carnivorous plants and can be grown in temperate, tropical and sub tropical climates outside in full sunlight, provided the plant has ample water.

As spring approaches flower scapes appear, rising as far as 30 cm (12 in) above the ground. Where light is stronger and conditions harsher the flower scape may be no taller than 15 cm (6 in). Each scape has 1-15 flowers, each of which has five white petals. To increase seed production on your adult plant, apply pollen from the stamen of one *Dionaea* to the stigma of another using a soft paint brush. Within two months the seeds will cluster inside the black withered petals. Plants need to be at least three years old before flowering occurs, by which time the rhizome is 1.5 cm (0.6 in) wide, and the plant has six or more traps. Flowers are no more than 1.5 cm (0.6 in) across, branching three times or more on the last 3 cm (1.2 in) of the scape. Each flower has its own short stalk [3 cm (1.2 in) long] which grow in spoke-like formations, the innermost flowers opening first, two at a time. Flowers remain open for about five days before they wither to brown pods that encase developing seeds which, when ripe, are small, black, shiny and pear-shaped.

Dionaea muscipula

Dionaeae muscipula

Dionaea muscipula

Common Name: Venus Fly Trap.
Size: In cultivation leaves will grow to 10 cm (4 in) or longer with six traps 10-25 mm (0.4-1 in) long, although variations do occur.
Climate: Subtropical, humid.
Habitat: Open grassy plains in damp sandy acidic soils in areas fringed or dotted with pine trees and frequented by fire.
Distribution: Coastal plain of North and South Carolina USA.
Flowering Time: During the wet season of late spring to early summer, or one to two months earlier in cultivation.
Colour: Leaves usually green, or with a dark burgundy tinge in summer; traps red to dark red on the inside and green on the outside.
Description: *Dionaea* is a perennial genus. It develops from a white-tissued base that grows longer with each new leaf. The leaf blade is usually 5-10 cm (2-4 in) long and 2-4 cm (0.75-1.5 in) wide over summer and 2-4 cm (0.75-1.5 in) long during winter. Each leaf blade has two hinged lobes 25 mm (1 in) long, looking like eyelids about to blink. The bristles on the edge of the rim are firm and about 4-8 mm long (0.16-0.32 in); the largest are at the crest, decreasing in size at each end, and pointing slightly inwards when the trap is open. Within and on either side of each lobe are three small trigger hairs, red or green to camouflage with the inside of the trap.

Plants grown in a glasshouse tend to produce fewer flowers on longer scapes (40 cm (16 in))—they wither quickly and can appear three or more weeks before those grown outside.

Leaves grow from the end of a long (1-6 cm (0.4-2.4 in) or more) thin white horizontal rhizome. Once the leaves are half grown the traps begin to lift away from the leaf and by leaf maturity the outer 'fangs' begin to pop out. The 'fangs' will form a cage as the trap begins to close, preventing insect escape. Within a few days of the fangs emerging the trap opens to an angle of about 60 degrees, and the trigger hairs bend upwards. The scene is set: the trap is in tension, awaiting prey.

DROSERA
'drŏss-ė-ră

Drosera belongs to the family Droseraceae and was classified by Carl Linaeus in 1753. *Drosera* comes from the Greek word *drosos* meaning dew, referring to the droplets of dew that glisten on the plants. The best time to see this effect is in the early morning as the fog lifts.

Of the more than 104 species of *Drosera* scattered throughout the world, over half are native to Australia. It is second only to *Utricularia* in terms of diversity and range amongst carnivorous plants. *Drosera* exists in almost all climate types, from the snow-covered alps of New Zealand to the warm dry plains of Australia, from the swamps of North America to the jungles of Borneo.

So diverse is this genus that one of the most useful ways of dividing it up is to observe the structure below the soil line, which is either tuberous or non-tuberous. Further, above the soil it can be erect, climbing, scrambling, fan-leafed, rosette or pygmy.

Drosera flowers usually last for one day and are self-fertilising. Each plant produces from one to 20 or more flowers; each flower has five petals and five stamens. Most plants produce many seeds (0.5–1 mm (0.002–0.009 in)) long while others, like the pygmy *Drosera*, produce gemmae and species such as *D. prolifera* produce plantlets on runners. The largest *Drosera* can grow up to 1.5 m (4.9 ft) high and the smallest is about 1 cm (0.4 in) in diameter.

Drosera are the easiest of all carnivorous plants to grow: the trick is to start with species that grow in or near your area, or those of similar climate and then to experiment with other species. Once this is achieved, you can produce hybrids by cross-pollination.

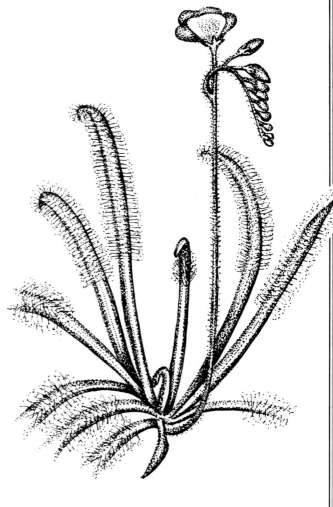

Drosera capensis

Drosera adelae

Common Name: Lance-leaved Sundew.
Size: 20 cm (8 in).
Climate: Tropical, humid.
Habitat: Creeks and banks in shaded areas; amongst short scrub in areas of high rainfall.
Distribution: Coast of northern Queensland, Australia.
Flowering Time: Spring, prior to monsoon, or mid-summer in cooler climates when under cultivation.
Colour: Leaves dark green in shade, burgundy in full sunlight; flowers red to apricot or greenish white.
Description: A non-tuberous herbaceous perennial with long narrow lanceolate leaves up to 20 cm (8 in) with a prominent main rib and sparse tentacles. Usually more than 20 flowers appear on each scape, 5 mm (0.2 in) across, they open one at a time. Plants that either die back or are pruned produce two or more new plants when conditions are right. Plantlets will grow out of drainage holes in the base of pots, and are ideally suited to terrariums.

Drosera aliciae

Drosera aliciae

Common Name: None.
Size: 5 cm (2 in) in diameter.
Climate: Subtropical to temperate, subhumid.
Habitat: Damp moss and wet sand, in areas where afternoon mist provides shelter from the hot sun.
Distribution: South Africa.
Flowering Time: Summer.
Colour: Bright red leaves; purple flowers.
Description: A non-tuberous rosette *Drosera* with spathulate leaves 2.5 cm (1 in) long and 7 mm (0.25 in) wide, with tentacles along almost the entire length. Older leaves remain on the plant, resulting in a mounded appearance. The scape is up to 40 cm (16 in) long and has 2–12 flowers, 2 cm (0.75 in) in diameter. Two or more flowers are often open at any time.

A prolific self-pollinating species, *D. aliciae* is ideally suited to a terrarium or peat garden, and its thick roots also make it ideal for root cuttings.

Drosera adelae

Drosera andersoniana

Common Name: Sturdy Sundew.
Size: Up to 25 cm (10 in) high.
Climate: Subtropical, arid.
Habitat: Loam soil near rocky outcrops.
Distribution: Inland of south western Australia.
Flowering Time: Late winter to early spring.
Colour: Plant dark red to brown; flowers white to pink or, occasionally, red.
Description: As with many erect *Drosera*, *D. andersoniana* has a basal rosette that emerges before the plant begins to climb. It has a yellow tuber 7 mm (0.25 in) across, and the leaves are 2 cm (0.75 in) long, and cup-shaped. The racemose inflorescence contains 3-15 flowers 1-1.5 cm (0.4-0.6 in) in diameter. It frequently grows with *D. peltata*.

Drosera binata

Drosera binata

Common Name: Forked Sundew.
Size: Varies, up to 30 cm (12 in).
Climate: Tropical, subtropical, temperate, humid to subhumid.
Habitat: Bogs and swamps in wet acid soil; both filtered light and full sunlight.
Distribution: New Zealand and eastern states of Australia.
Flowering Time: Late spring to early summer.
Colour: Leaves green with burgundy tentacles, sometimes dark burgundy; in full sunlight flowers white or (sometimes) pale pink.
Description: A non-tuberous erect Sundew. Many varieties, but all have leaf blades that fork into a 'Y' shape one or many times. The leaves are slender and covered densely in tentacles. Inflorescence is branched with 15-30 flowers.

This is one of the few *Drosera* that looks particularly attractive in a sphagnum hanging basket. It is one of the easiest and quickest species to grow.

Varieties: *Drosera binata* 'T' form, *D. binata* 'dichotoma', *D. binata* 'multifida'.

Drosera andersoniana

Drosera brevifolia

Drosera brevifolia

Common Name: Dwarf Sundew.
Size: 2 cm (0.75 in) in diameter.
Climate: Tropical to subtropical, humid to subhumid.
Habitat: Sandy soils.
Distribution: Mexico, USA.
Flowering Time: Spring.
Colour: Leaves green to red; flowers pink or white.
Description: One of the smallest non-tuberous *Drosera*, this species has a rosette with leaves up to 1 cm (0.4 in) long and 4–8 mm (0.2–0.4 in) wide, and petioles that are tiny or even non-existent. A glandular pubescent scape up to 8 cm (3 in) high bears between one and eight self-pollinating flowers, each 1.5 cm (0.6 in) in diameter. It tends to exist in drier habitats than other American *Drosera*, and is often both found and confused with *D. capillaris*.

Drosera capensis

Drosera capensis

Common Name: Cape Sundew.
Size: Up to 15 cm (6 in).
Climate: Temperate, humid to subhumid.
Habitat: Bogs, marshes, wet grasslands; in peaty soil in both full sunlight and filtered light.
Distribution: South Africa (the Cape).
Flowering Time: Spring.
Colour: Leaves green with burgundy to brown tentacles; flowers bright pink.
Description: A non-tuberous erect Sundew. Slender flat leaves form a single petiole growing from the base of the plant. The ends of the leaves are rounded and curl around the prey. Up to 20 flowers open one at a time, beginning at the base. Old plants produce a long black trunk with green foliage at the top, looking like a miniature palm tree.

Probably the easiest to grow and most prolific of all *Drosera*, *D. capensis* is able to withstand both heat and moderate cold. Root cuttings of 1.5–2.5 cm (0.6–1 in) form buds in two to three weeks under ideal conditions. *D. capensis*, as all other *Drosera*, benefits from a light spray with liquid fertiliser on the leaves once a month. The tentacles will fold in for a few hours, but will soon fold back again.

Drosera capillaris

Drosera cistiflora
Common Name: Pink Sundew.
Size: Usually 3 cm (1.25 in) in diameter, but there is also a large form that is up to 8 cm (3 in) in diameter.
Climate: Tropical–temperate, humid.
Habitat: Damp soil.
Distribution: East coast of North America; Guyana, Venezuela.
Flowering Time: Summer.
Colour: Leaves green to bright red; flowers pink or white.
Description: One of the most prolific and easiest to grow *Drosera*, this is a non-tuberous rosette species with spathulate leaves up to 1.5 cm (0.5 in) long and 4 mm (0.2 in) wide. Up to 12 self-pollinating flowers, 6 mm (0.25 in) in diameter, grow on two or more scapes up to 15 cm (6 in) high.

Ideal for a peat garden or terrarium, *D. capillaris* often grows with *D. brevifolia* and *Sarracenia oreophylla*.

Drosera capillaris
Common Name: None.
Size: Up to 30 cm (12 in).
Climate: Temperate, humid to subhumid.
Habitat: Areas where the soil is wet during autumn, winter and spring and dry during summer.
Distribution: South Africa (the Cape).
Flowering Time: Spring.
Colour: Leaves green but appear red/burgundy due to the covering of tentacles; flowers usually pale pink but can be pink, purple, white or yellow.
Description: A non-tuberous, erect, very pretty *Drosera* with fine leaves up to 10 cm (4 in) long. Leaves are rounded at the ends and curl around the prey. Inflorescence usually single-flowered, self-pollinating, occasionally up to three flowers, rarely up to six. Usually dormant in the dry summer, emerges from a rhizome. In cultivation, where the soil does not dry out, *D. cistiflora* grows all year.

Drosera cistiflora

Drosera dichrosepala

Drosera ericksonae

Drosera ericksonae

COMMON NAME: Bright Sundew.
SIZE: 1 cm (0.4 in) in diameter.
CLIMATE: Subtropical, subhumid.
HABITAT: Wet sandy soil in full sunlight.
DISTRIBUTION: South western Australia.
FLOWERING TIME: Spring to summer.
COLOUR: Light to dark pink flower has a green centre with orange-yellow surrounds.
DESCRIPTION: Previously known as *D. omissa*, *D. ericksonae* is a non-tuberous pygmy *Drosera* that produces gemmae. Spathulate leaves are green at the base and orange-yellow towards the tip, measuring 3 mm (0.12 in) across. The stem on this pygmy *Drosera* species is short, while the four scapes are 4-7 cm (1.75-2.75 in) long. Each scape has six or more flowers, each measuring 9 mm (0.35 in) across. The plant produces over 1000 gemmae, each 1 mm (0.04 in) across, each season. Reproduces by seed each season.

Drosera dichrosepala

COMMON NAME: Rusty Sundew.
SIZE: 5 cm (2 in) high, 6 mm (0.24 in) across.
CLIMATE: Temperate, subhumid, dry summer.
HABITAT: Damp clay soil amongst scrub.
DISTRIBUTION: South west Australia.
FLOWERING TIME: Spring.
COLOUR: Leaves pale green with red to orange tips; flowers creamy white.
DESCRIPTION: A pygmy *Drosera* with a rosette that measures 6 mm (0.24 in) across. Old leaves wither but remain on the plant and new leaves—oblong to spathulate, 3 mm (0.12 in) long—emerge from the centre, so that a small dark brown to black trunk develops up to 5 cm (2 in) high. One or two scapes, 7-20 mm (0.25-0.5 in) long with two to seven scented flowers, each 6-10 mm (0.24-0.4 in) across.

Drosera ericksonae

Drosera erythrorhiza

COMMON NAME: Redink Sundew.
SIZE: Up to 10 cm (4 in) in diameter.
CLIMATE: Subtropical to temperate, subhumid.
HABITAT: Open scrub in sandy peat soil.
DISTRIBUTION: South west Western Australia.
FLOWERING TIME: Autumn.
COLOUR: Leaves bright green with red tentacles, or reddish in older plants and those in full sunlight; white flowers.
DESCRIPTION: A tuberous rosette *Drosera*, the round 2.5 cm (1 in) leaves have a central depressed valley running from base to tip. Up to 50 flowers. One of the largest rosette Sundews, it produces smaller secondary tubers, as do many of the tuberous *Drosera*. If grown in a pot these secondary tubers usually touch the sides.

Flowering increases after fire. Burning dry leaves on the surface during summer will promote flowering before leaves appear, but it can also cause *D. erythrorhiza* to exhaust itself and die.

There are four sub-species of *D. erythrorhiza*:
D. erythrorhiza ssp. *collina*: the largest species, has 8-9 leaves, entire plant 12 cm (4.75 in) in diameter
D. erythrorhiza ssp. *magna*: slightly smaller with 6 leaves
D. erythrorhiza ssp. *squamosa*: up to 8 leaves and 6 cm (2.5 in) in diameter
D. erythrorhiza ssp. *erythrorhiza*: similar size to *squamosa* but with only 3-5 leaves.

Drosera filiformis

Drosera erythrorhiza

Drosera filiformis

COMMON NAME: Threadleaf Sundew, Dew Thread.
SIZE: Up to 30 cm (12 in) for form *filiformis* and almost double that for var. *tracyi*.
CLIMATE: Temperate, humid to subhumid.
HABITAT: Swampy clearings, moist sandy soil.
DISTRIBUTION: USA, east coast and south east.
FLOWERING TIME: Spring.
COLOUR: Leaves green with red tentacles; flowers pink.
DESCRIPTION: A non-tubereous, erect species, it has a very short stem with fine long leaves that uncurl like fern fronds. These sticky tentacles capture many insects but the leaves do not curl around the prey. Each plant produces up to three scapes, with one to 25 self fertilising flowers, on each scape.

Produces hibernaculum, during winter, at which time the plant should be kept drier. As spring approaches separate the hibernaculum, repot and increase watering.

VARIETIES: Two recognised types exist, *D. filiformis* form *filiformis* and *D. filiformis* var. *tracyi*.

Drosera gigantea

Common Name: Giant Sundew.
Size: 1 m (3.3 ft).
Climate: Subtropical, subhumid.
Habitat: Bogs, creek beds and sandy soil; often sits in water during winter.
Distribution: South west Western Australia.
Flowering Time: Spring to summer.
Colour: Leaves mustard yellow on a green stem; flowers white.
Description: The tallest of the tuberous *Drosera*, this is a tall free-standing plant that reaches a height of 1 m (3.3 ft). It is like a tree, branching every inch with shield-shaped leaves. The small red tuber may extend to 75 cm (30 in) below the surface in dry soil, or 30 cm (12 in) where the soil is damper. Flowers are numerous; 200 or more is quite common on a single plant. A difficult species to maintain year after year, it appears in early winter, then turns black and dries out in mid-summer. As with other tuberous *Drosera*, soil must be allowed to dry out during summer otherwise the tuber will rot.

Drosera glanduligera

Drosera glanduligera

Common Name: Common Scarlet Sundew.
Size: 1-2 cm (0.4-0.75 in) in diameter.
Climate: Temperate, subhumid.
Habitat: Varied—sandy clay soil, damp scrub, banks of creeks.
Distribution: Southern and western Australia.
Flowering Time: Spring to beginning of summer.
Colour: Leaves bright yellow, sometimes yellow-green; flowers red-orange with black centres.
Description: An annual and one of the few Australian non-tuberous rosette *Drosera* that die off totally in their natural habitat, to return the following year from the hundreds of seeds each plant produces. Almost round, 12 mm (0.5 in) diameter leaves form a convex rosette. Sparse red tentacles. Up to 20 flowers on 6 cm (2.5 in) scapes.

Drosera gigantea

Drosera hamiltonii

Common Name: Rosy Sundew.
Size: 4 cm (1.5 in) in diameter.
Climate: Subtropical to temperate, semi humid.
Habitat: Swamps and edges of streams.
Distribution: South western Australia.
Flowering Time: Late spring, early summer.
Colour: Leaves green to reddish; flowers pink to purple.
Description: A very prolific species when grown in a bog garden, this non-tuberous *Drosera* has fleshy spathulate leaves up to 2 cm (0.5 in) long and covered with dark red hairs. A single scape up to 30 cm (12 in) high has 5-20 flowers, each 2 cm (0.75 in) in diameter. This fairly rare species is often found near *Cephalotus follicularis*.

Drosera hamiltonii

Drosera intermedia

Drosera intermedia

Common Name: None.
Size: 6 cm (2.5 in) in diameter.
Climate: Temperate–tropical; humid–sub humid.
Habitat: Wet areas bordering swamps and streams.
Distribution: Eastern side of North America; Europe, Guyana.
Flowering Time: Summer.
Colour: Leaves brownish green, reddish at the tip; flowers white.
Description: A non-tuberous rosette *Drosera* with oblong leaves up to 3 cm (1.25 in) long and 0.5 cm (0.2 in) wide. Leaves elongate in extremely wet conditions, and this *Drosera* produces a stem that can reach up to 20 cm (8 in) high. Up to 20 self-pollinating flowers, 6 mm (0.25 in) in diameter, appear on a single scape up to 15 cm (6 in) high.

During winter *D. intermedia* dies back to a resting bud that can be removed and transplanted in early spring. It is often found with *D. rotundifolia*.

Drosera leucoblasta

Common Name: Orange Sundew.
Size: 1-1.5 cm (0.4-0.6 in) in diameter.
Climate: Subtropical, subhumid.
Habitat: Stony or clay soil often moist during winter.
Distribution: South west Western Australia.
Flowering Time: Spring.
Colour: Leaves yellow with red centres, with translucent droplets on the tentacles. In full sunlight the lamina is dark green and petiole red; flowers red to metallic orange.
Description: A pygmy *Drosera* with a rosette of round leaves, each 2 mm (0.08 in) in diameter. Scape up to 10 cm (4 in) with up to seven flowers.

As with many pygmy *Drosera*, *D. leucoblasta* produces gemmae, which in turn produce mature plants after three months.

Drosera leucoblasta

Drosera linearis

Common Name: None.
Size: 5 cm (2 in) high.
Climate: Temperate to boreal; sub-humid.
Habitat: Full sunlight in alkaline bogs.
Distribution: Northern USA and Canada.
Flowering Time: Summer.
Colour: Leaves, green, flowers white.
Description: A non-tuberous *Drosera* with erect linear leaves up to 5 cm (2 in) long and 2 mm (0.08 in) wide. Six or more self-pollinating flowers 1.2 cm (0.4 in) in diameter on a scape up to 10 cm (4 in) high. Forms winter resting buds, which can be removed in early spring and transplanted. Is totally covered by water for short periods during the wet season.

Ideal in a peat garden and bowl arrangements, it is often found with *Sarracenia purpurea*, *D. rotundifolia* and *D. anglica*.

Drosera linearis

Drosera macrantha ssp *planchonii*

Common Name: Climbing Sundew.
Size: Up to 60 cm (24 in) high.
Climate: Temperate, subhumid.
Habitat: Sandy soil, open scrub.
Distribution: Central, south, south eastern and western Australia.
Flowering Time: Spring to summer.
Colour: Red-green leaves on a green stem; flowers white.
Description: A tuberous *Drosera* with a climbing stem and very small, scale-like, lower leaves. The cup-shaped leaves of the climbing stem, in contrast, are 5 mm (0.2 in) in diameter with spiky red tentacles. Flowers have five petals, are 2–3 cm (0.8–1.2 in) wide and, as with other climbing *Drosera*, up to five flowers appear at the tip of the plant.

Drosera macrantha ssp *planchonii*

Drosera menziesii ssp *menziesii*

Drosera menziesii ssp *menziesii*

Common Name: None.
Size: Up to 35 cm (14 in).
Climate: Subtropical, humid.
Habitat: Varied—swamps, scrub, wet sandy or clay soil.
Distribution: Western Australia.
Flowering Time: Spring.
Colour: Flowers white or pink to red; leaves red.
Description: A scrambling tuberous *Drosera* that relies on adjacent plants for support. Round, cup-shaped leaves, 4 mm (0.16 in) in diameter are bright red and appear on a delicate fine red stem. Up to six flowers.
Varieties: *D. menziesi* ssp *menziesi*; *D. menziesii* ssp *pencillaris*; *D. menziesii* ssp *thysanosepalia*; *D. menziesii* ssp *basifolia*.

Drosera microphylla

Drosera microphylla

Common Name: Purple Rainbow.
Size: Up to 40 cm (16 in) tall.
Climate: Subtropical to temperate, semi-humid.
Habitat: Damp sandy soils.
Distribution: South western Australia.
Flowering Time: Early spring.
Colour: Leaves green; flowers pink to dark red.
Description: An erect tuberous *Drosera* with small (0.5 cm (0.2 in)) leaves, usually scattered singly along the stem, but sometimes forming groups of three. Inflorescence is a terminal panicle, sometimes single-flowered, but usually up to ten flowers. Flowers are 1 cm (0.4 in) wide with prominent bright green sepals. Tuber is dark red, and 8 mm (0.3 in) in diameter.

Drosera modesta

Common Name: Modest Rainbow.
Size: Up to 80 cm (32 in) tall.
Climate: Temperate to subtropical, subhumid.
Habitat: In the shade in clay soil in Jarrah forests; in sunlight in swamp areas in sandy soil.
Distribution: South western Australia.
Flowering Time: Spring to late spring.
Colour: Leaves greenish yellow; flowers white.
Description: A scrambling tuberous *Drosera* that grows during the wet winter. The tuber is reddish brown and 6 mm (0.25 in) in diameter, and the scape a terminal panicle with up to 15 flowers, each 2 cm (0.75 in) across. Cup-shaped leaves are 3 mm (0.12 in) across with two pointed lobes; they grow in groups of three, with the largest providing support by attaching itself to other objects, usually other plants. *D. modesta* often grows alongside *D. pallida*, a taller thinner species.

Drosera paleacea

Drosera paleacea

Common Name: Dwarf Sundew.
Size: 1 cm (0.4 in) in diameter.
Climate: Subtropical to temperate, semi-humid.
Habitat: Sandy soils near swamps.
Distribution: South western Australia.
Flowering Time: Late spring, early summer.
Colour: Leaves green; flowers white.
Description: A non-tuberous pygmy *Drosera* with a compact rosette of 20–30 leaves 4 mm (0.16 in) long, and a lamina almost-round and 2–3 mm (0.080–.12 in) in diameter. The previous year's leaves wither but remain on the plant, forming a short stem up to 2 cm (0.75 in) high. The one to four scapes are up to 5 cm (2 in) tall with 25 or more small flowers, each 0.5 cm (0.2 in) in diameter. Two or three very long (30 cm (12 in)) roots ensure that during the hot dry summer the dormant bud does not wither and die.

Drosera modesta

Drosera parvula

Common Name: Cone Sundew.
Size: 1 cm (0.4 in) in diameter.
Climate: Subtropical, subhumid to humid.
Habitat: Sandy soil in open scrub.
Distribution: South western Australia.
Flowering Time: Spring.
Colour: The flower has a greenish centre with red spots surrounded by a white to red ring; leaves green, tipped red.
Description: This is a non-tuberous pygmy *Drosera* previously known as *D. androsacea*. The leaves from one year remain to produce a stem 1 cm (0.4 in) long the next. The rosette leaves have a red tip 2-3 mm (0.08-0.12 in) across and the single scape is up to 4 cm (1.6 in) long with four to five flowers, each about 1 cm (0.4 in) in diameter. As with many pygmy *Drosera*, it produces gemmae.

Drosera peltata ssp *auriculata*

Drosera peltata ssp auriculata

Common Name: The Aboriginal name for *D. peltata* ssp *auriculata* is *Errienellam* meaning 'entrapping hair'.
Size: Up to 70 cm (28 in), usually 10-30 cm (4-12 in).
Climate: Temperate, humid and subhumid.
Habitat: Bogs or permanently wet grasslands.
Distribution: East coast of Australia, South Australia, Tasmania and New Zealand.
Flowering Time: Spring to early summer.
Colour: Leaves yellow to green with burgundy tentacles; flowers white or (sometimes) pink.
Description: A tuberous *Drosera* similar to, and often confused with, *D. peltata*. Has spotted sepals and an erect stem with few branches, usually on the upper part of the stem. Lower leaves form a small basal rosette, while upper leaves are peltate on a slender petiole. Inflorescence a raceme with 5-10 flowers. Dormant during the hot dry summer.

Drosera parvula

Drosera peltata

Common Name: Pale Sundew.
Climate: Up to 25 cm (10 in).
Climate: Subtropical, temperate, subhumid, dry summer.
Habitat: Marshes, damp peat or clay soil, open scrub.
Distribution: Australia, China, India, Japan, Malaysia, and South East Asia.
Flowering Time: Spring.
Colour: Leaves yellow-green; flowers white or pale pink.
Description: The most widely distributed of all tuberous *Drosera*, *D. peltata* has a basal rosette of 3 mm (0.1 in) long leaves with fine tentacles, out of which it climbs to an erect Sundew. The leaves of the upper stem are also 3 mm (0.1 in) in length and are cup-shaped on a fine petiole. Inflorescence 2-10 flowered. Unlike many other *Drosera*, *D. peltata*'s flowers open during the middle of the day; however like other tuberous *Drosera* little success has been achieved with leaf cuttings, and no known hybrids exist. Does not usually last more than one season in a terrarium.

Drosera peltata

Drosera prolifera

Common Name: Trailing Sundew.
Size: Up to 10 cm (4 in) in diameter.
Climate: Tropical, humid.
Habitat: Tropical rain forests under filtered light near creek banks.
Distribution: Queensland, Australia.
Flowering Time: All year.
Colour: Leaves bright green; flowers pink, white or red.
Description: *D. prolifera* is non-tuberous, and has a rosette with up to 15 kidney-shaped leaves on a long petiole. Produces one or two long (18 cm (7 in)) trailing scapes, each with four to six flowers. This is the only *Drosera* that sends out radial runners, which take root and produce new plants—one specimen can produce up to 100 new plantlets each season.

Ideally suited to terrariums.

Drosera pulchella

Drosera pulchella

Common Name: Pretty Sundew.
Size: 1–2 cm (0.4–0.8 in) diameter.
Climate: Temperate, subhumid.
Habitat: Swamps and wet peat, damp sandy soil.
Distribution: South west of Western Australia.
Flowering Time: Summer.
Colour: Leaves green with short red tentacles, looking like a red circle with a green centre; flowers have a metallic sheen in white, pink, orange or dark pink to red.
Description: A pygmy *Drosera*. Tentacled leaves are round on broad petioles set in a convex rosette. Inflorescence, two to three scapes up to 5 cm (2 in) long with several 9 mm (0.35 in) flowers, each with five petals.

As with other pygmy *Drosera*, gemmae are produced in response to dramatic changes in condition, particularly of temperature or rainfall.

Drosera prolifera

Drosera pygmaea

Drosera pygmaea

Common Name: Pygmy Sundew.
Size: About 1 cm (0.4 in) in diameter.
Climate: Temperate to subtropical, humid to subhumid.
Habitat: Swamps, bogs, damp peat and sandy soil.
Distribution: Eastern Australia and New Zealand.
Flowering Time: Spring–summer.
Colour: Leaves red in full sun, green in the shade; flowers white.
Description: A non-tuberous perennial rosette *Drosera* with 2 mm (0.08 in) long leaves on slightly longer flat petioles around a tiny stem. It has a single very small flower less than 6 mm (0.24 in) on a thread-like scape; unusually for *Drosera*, the flower has four petals or, sometimes, the usual five. *D. pygmaea*, like other pygmy *Drosera*, produces gemmae, as small as 1 mm (0.04 in) long in their centre. One of the easiest *Drosera* to grow, ideally suited to a terrarium or bench arrangement where it will soon carpet the whole area.

Drosera pygmaea

Drosera pygmaea

Drosera spathulata

Common Name: Spoon-leaf Sundew.
Size: Up to 5 cm (2 in) in diameter.
Climate: Temperate, subtropical.
Habitat: In damp soil in wetlands and heath.
Distribution: Eatern Australia, New Zealand, Japan, China, Borneo, the Philippines, Taiwan, Hong Kong.
Flowering Time: Late spring to summer; throughout the year in southern areas and under cultivation.
Colour: Leaves green, or red in full sunlight; bright red tentacles the length of the leaf blade; flowers pink or white.
Description: A non-tuberous rosette *Drosera* with 20 or so spoon-shaped leaves that curve inwards slightly in their upper section. The petioles are up to 1 cm (0.4 in) long, covered in bristly tentacles. Two flower scapes are up to 20 cm (8 in) long with up to 15 flowers, each 12 mm (0.5 in) across.

Drosera spathulata

Drosera stolonifera

Drosera subhirtella

Common Name: Sunny Rainbow.
Size: Up to 50 cm (20 in) high.
Climate: Temperate to subtropical, subhumid.
Habitat: Peat bogs, often found with *Cephalotus* in moist white sand.
Distribution: South eastern corner of Western Australia.
Flowering Time: Late winter to spring.
Colour: Leaves reddish; flowers and tuber bright yellow.
Description: A tuberous *Drosera* with a slender climbing or trailing stem, it usually requires other plants for support. Lower leaves are little more than scales while upper leaves, which grow in threes, are round and up to 6 cm (2.4 in) in diameter, on fine petioles. Inflorescence a terminal panicle, with up to 20 flowers, each about 2 cm (0.8 in) across.
Varieties: *D. subhirtella* var. *moorei* (slightly smaller).

Drosera subhirtella

Drosera stolonifera ssp *compacta*

Common Name: Leafy Sundew.
Size: 12 cm (5 in) high.
Climate: Temperate, subhumid.
Distribution: South western Australia.
Flowering Time: Mid-winter to spring.
Colour: Leaves yellow to red; flowers white.
Description: A stout, erect, tuberous *Drosera* with a 1.5 cm (0.5 in) basal rosette, it branches at the base to produce four or more stems 10 cm (4 in) long. Each stem branches in turn, with fan-shaped leaves 1 cm (0.4 in) wide attached. A single flower scape rises from the centre of the basal rosette to a height of 5 cm (2 in) and has 20 or more flowers.

D. stolonifera ssp *compacta* is one of four subspecies, the others being: ssp *rupicola*, ssp *humilis* and ssp *stolonifera*.

Drosera whittakeri

COMMON NAME: Scented Sundew.
SIZE: 6 cm (2.4 in) in diameter.
CLIMATE: Temperate.
HABITAT: Damp open spaces and in humus in a clay, sandy soil.
DISTRIBUTION: South eastern Australia.
FLOWERING TIME: Winter to spring.
COLOUR: Leaves bright green, or reddish with red tentacles in full sunlight; large white flowers.
DESCRIPTION: A tuberous rosette *Drosera* with round or spoon-shaped sessile leaves 1 cm (0.4 in) in diameter that grow in a flat rosette from a short stem. Several short (2 cm (0.8 in)) scapes, each with a single scented flower 3 mm (0.1 in) wide. Leaf cuttings have proved unsuccessful.

Early European settlers used the plants to make red ink.
VARIETY: *D. whittakeri* var *praefolia*.

Drosera whittakeri.

Drosera helodes (hybrid: publication in preparation).

Drosophyllum
drŏs-ō-fī-lim

Drosophyllum is a monotypic genus of the Droseraceae family, found below 200 m (656 ft) above sea level on the coasts of Portugal, south west Spain and the northern tip of Morocco. The name comes from the Greek, *drosos* meaning dew and *phyllon* meaning leaf. The first detailed account of *Drosophyllum* was published by H.F. Link in 1805, though the existence of this plant was recorded as early as the seventeenth century.

Drosophyllum is superficially similar to many *Drosera*, but is in fact quite different. The plant is easy to grow and maintain once established.

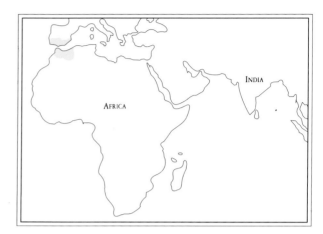

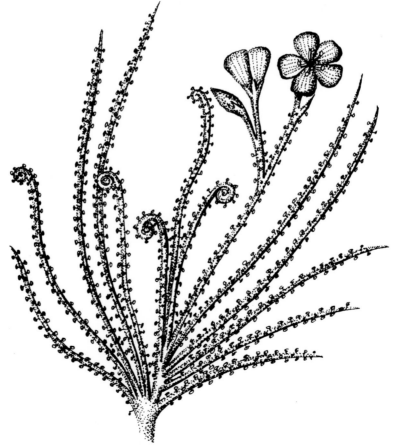

Drosophyllum lusitanicum

Drosophyllum lusitanicum

Drosophyllum lusitanicum

Common Name: Portuguese Sundew.
Size: Up to 30 cm (12 in) high.
Climate: Temperate, semihumid.
Habitat: In dry soil on coastal hills in areas full of small quartz-like stones that provide rapid drainage, or in coastal regions on mountains amongst short fir trees (*Pinus pinaster*).
Distribution: Portugal, Spain, Morocco.
Flowering Time: Early summer.
Colour: Leaves light green; inflorescence bright sulphur yellow.
Description: *D. lusitanicum* has a woody base and 100 or more thin leaves supported by a long fibrous root that penetrates deep into the cracks between the rocks to absorb any seepage. Thin leaves have short tentacles that unravel outwards as they emerge and secrete a viscid liquid. Unlike *Drosera*, the tentacle-like leaves do not move, but rather the insect smothers in mucilage. Also unlike *Drosera* the leaves curl 'outwards' rather than 'inwards'.

The flowers are 3 cm (1.2 in) in diameter, stay open for one day and need pollinating to self-seed.

This stocky plant does not normally live more than six years. Slowly the plant turns brown from the base up, and dies.

Drosophyllum lusitanicum

Genlisea
'djĕn-,lĭss-ē-à

Genlisea, a member of the family Lentibulariaceae, has 15 known species native to tropical Africa, South Africa, South America and Madagascar. Auguste de Saint-Hilaire described the first Genlisea in Brazil in 1833, and named it after Stephanie de Genlis. *Genlisea* is usually found at altitudes of 1000–2000 m (3200–6400 ft) above sea level, although one species is found on Brazil's Mount Roraima at 2800 m (9000 ft), in moist sandy peat soil.

Genlisea is a rootless, terrestrial, perennial plant with numerous green linear to spathulate leaves. They are up to 20 cm (8 in) long and 9 mm (0.35 in) wide, and appear above soil level giving little indication that below is a second type of 'leaf' that catches small creatures in a trap similar to that used for lobsters or, as Darwin commented, eels.

There are up to 10 flowers on a single scape 30 cm (12 in) long. The flowers measure up to 13 cm (5 in) long and range in colour from purple, deep blue, and pale purple to yellow. The seeds are quite large, oval shaped and up to 1 cm (0.4 in) across, appearing almost triangular in cross section.

There are few plants under cultivation, but those that are, are best grown in a tropical environment, under filtered sunlight.

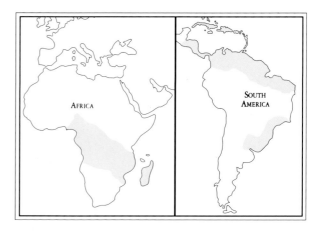

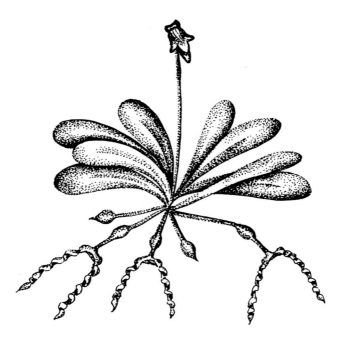

Genlisea hispidula

Genlisea guianensis

Common Name: None.
Size: Leaves 1–10 cm (0.4–4 in) long.
Climate: Tropical, humid, lowland.
Habitat: Wet savannah and on rocks by streams.
Distribution: Venezuela, Guyana, at altitudes of 500–1000 m (1600–3200 ft).
Colour: Leaves green; flowers pale violet to purple.
Description: Typical of the genus except that leaves are lanceolate, up to 20 cm (8 in) long and 1–2 mm (0.04–0.08 in) wide. Flower scape 18 cm (7 in) long with 2 to 4 flowers, each 7 mm (0.3 in) wide. Plantlets often develop at the branching of the flower scape.

Genlisea guianensis

Genlisea guianensis

Genlisea repens

Common Name: None.
Size: Leaves 0.5–4 cm (0.2–1.5 in) long.
Climate: Tropical, highland, humid.
Habitat: By watercourses and in swampy savannahs, at altitudes of 1700–2100 m (5500–6800 ft).
Distribution: Brazil, Guyana, Paraguay, Suriname, Venezuela.
Flowering Time: Spring to summer.
Colour: Leaves green; flowers yellow.
Description: A terrestrial herb with densely packed green leaves, obovate to spathulate, 1–4 cm (0.4–1.5 in) long and 1–1.5 mm (0.04–0.06 in) wide. The flower scape is 18 cm (7 in) long and carries 2–4 flowers, each 7 mm (0.3 in) wide.

Genlisea repens

Heliamphora
hē-li-ăm-´for-à

Heliamphora is a rare genus, with five species. It is part of the Sarraceniaceae family, and *Heliamphora* are commonly called Sun or Marsh Pitchers. The name *Heliamphora* comes from the Greek *helios*, meaning marsh and *amphora*, meaning pitcher. The first species to be described was *H. nutans*, found by Robert Schomburg on Mount Roraima in Guyana and sent to George Bentham who published a detailed description in 1840.

All species have similar, tubular-shaped, green to red pitchers with a small overhanging cap, except for *H. heterodoxa* var. *exappendiculata*, which has no cap. Pitchers are attached to a stem at the base, connecting the pitcher to the rhizome. The stem can be as short as a few cm or as long as 3.5 m (11.5 ft). *Heliamphora* are usually found in clumps, and all species have a branching rhizome with pitchers developing in rosette form. Pitchers can be as small as 4 cm (1.6 in), where only the water-filled opening is observed above the ground or, in more exposed areas where the soil is very moist, can reach up to 50 cm (20 in) in height.

Interestingly, all pitchers on a single plant have a similar water level—always below the rim—despite their different heights and heavy rain. This water level is maintained by a slit or seam in the pitcher that provides an overflow outlet. Without this slit, pitchers could become top heavy and fall over and nourishment would be washed away by the constant rains. It has been suggested that *Heliamphora* are the ancestors of *Sarracenia* with the overhanging cap becoming a lid and the slit disappearing.

Heliamphora have a racemose inflorescence with several slightly or non-scented flowers on a scape up to 60 cm (2 ft) high. It is believed that pollination is caused by the wing vibration of bees causing a shower of pollen. Seeds are numerous, irregularly oval, winged and 1 mm (0.04 in) long. It is easily grown outdoors in tropical climates and in a glasshouse in cooler climates. High humidity is essential, especially when temperatures are high. Avoid temperatures above 30°C (86°F).

Heliamphora are native to the tepuis (high sandstone plateaus believed to be over 1600 million years old) in Venezuela and Brazil. These tepuis are at altitudes between 1000 and 3000 m (3200–9800 ft). The area is a marshy savannah with bright sunlight, filtered, in places, by the mist.

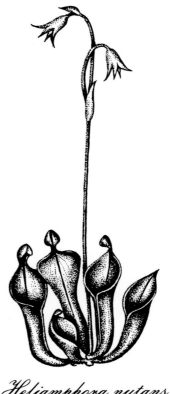

Heliamphora nutans

Heliamphora heterodoxa

Common Name: None.
Size: Up to 42 cm (16.5 in).
Climate: Temperate—tropical, humid.
Habitat: Open swampy savannah and mossy rock exposures at elevations of 1000–3000 m (3200–9800 ft).
Distribution: Torono Tepui, Chimanta Tepui, Auyan Tepui and Ptari Tepui, Venezuela.
Flowering Time: Early to mid-winter.
Colour: Pitchers olive green with a thin red line on the rim, defining its shape—older pitchers have a red cap, younger ones a green cap; flowers pink and white.
Description: Pitchers can achieve a height of 42 cm (16.5 in). The hairy zone within the pitcher is 6–20 cm (2.4–8 in) for this species as for *H. tatei* and *H. tatei* var. *neblinae* (the hairy zone is larger for other *Heliamphora*). The pitcher is distinctly waisted a third of the way up its length. This is one of the easier species to grow.
Varieties: *Heliamphora heterodoxa* var. *exappendiculata*, *H. heterodoxa* var. *glabra*, *H. heterodoxa* var. *heterodoxa*.

Heliamphora heterodoxa

Heliamphora heterodoxa

Heliamphora heterodoxa

Heliamphora minor

Heliamphora minor

Common Name: None.
Size: Up to 8 cm (3 in).
Climate: Temperate to tropical, humid; high altitudes.
Distribution: Auyan Tepui and Chimanta Tepui, Venezuela.
Habitat: Between altitudes of 1800–2000 m (5900–6500 feet), in swampy and rocky savannah and bordering waterfalls and rivulets.
Flowering Time: Spring to summer.
Colour: Pitchers green, or red to burgundy in full sunlight; small overhanging lid red to burgundy.
Description: The smallest of the *Heliamphora*, *H. minor*'s cone-shaped pitchers grow to 8 cm (3 in) tall and 1.5 cm (0.6 in) in diameter, although pitchers have been found measuring 10–15 cm (4–6 in) in height. Emerging pitchers are covered with short, fine, white hairs, and in full sunlight they look like shiny pink slippers with bright red toes. When the pitcher first opens, the rim is red and the interior is green; as it flares open the whole interior darkens to burgundy and the fine outside hairs remain only on the central seam and along the rim of the pitcher. This species is so abundant on Auyan Tepui that it is difficult to avoid stepping on them—they grow everywhere, in patches up to 1 m (3.3 ft) wide.

Heliamphora tatei

Common Name: None.
Size: Up to 5 m (16 ft) in open areas and up to 1 m (3.3 ft) in marshy areas.
Climate: Temperate to tropical, humid.
Distribution: Mount Duida, Mount Marahuaca and Mount Huachamacari in Venezuela at altitudes of 2000 m (6400 ft).
Flowering Time: Winter.
Colour: Flowers are white to red.
Description: The densely packed pitchers are elongated and slightly pinched in in the middle, with a cap 3 cm (12 in) wide. The stem that bears the pitchers can be as long as 3 m (9.9 ft). If the area is exposed and drier, only one pitcher may appear on each plant.
Varieties: *H. tatei* var. *macdonaldae*, *H. tatei* var. *neblinae*.

Heliamphora tatei var. *neblinae*

Heliamphora tatei

Heliamphora tatei var. neblinae

Common Name: None.
Size: Up to 25 cm (10 in).
Climate: Temperate to tropical, humid.
Habitat: Swampy savannah, in open woodland.
Distribution: Cerro de la Neblina (Mountain of Clouds) on the Venezuela and Brazilian border, at altitudes of 1300-2800 m (4200-9100 ft).
Flowering Time: Winter.
Colour: Pitchers yellow green; flowers white, turning pink with age.
Description: Pitchers are 15-25 cm (6-10 in) high, with a bright red rib running the inside length. The cap is bright red on the underside and green on top, and the wing in the front is wide at the base, to provide support, and narrower at the top. Over 100 flies have been found in each pitcher, and in some mosquitoes breed.

Nepenthes
nē-pĕn-thēz

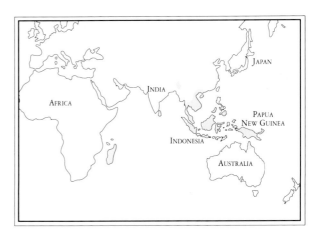

Nepenthes has at least 70 species. *Nepenthes* is derived from the Greek *nepenthe*, the drug Helen mixed with wine in Homer's *Odyssey*, essentially meaning 'without care'. The first *Nepenthes* known to Europeans was published using the Malagasy name 'Amramitico' by Flacourt and later changed to *Nepenthes* by Carl Linnaeus in 1753. Found in the humid tropics of Madagascar, the Seychelles, Sri Lanka, the northern tip of Australia, Singapore and Malaysia, Indonesia, Papua New Guinea, Thailand, Vietnam, India, Philippines and New Caledonia, the genus is more or less bounded either side of the equator by the tropics of Cancer and Capricorn.

Nepenthes are divided into two main groups, the highland and the lowland species. Highland species are found at altitudes greater than 1000 m (3200 ft) above sea level and experience relatively cool nights, often as low as 5°C (41°F)—difference in temperature between summer and winter is slight, the greatest variance being that between day and night. Highland plants can be cultivated in full sunlight, as long as it is in association with constant misting, provided by foggers. The more common lowland species, growing below 1000 m (3200 ft), exist in a warmer climate, but only experience temperatures as low as 15°C (59°F). *N. alata* is one of the few species found in both highland and lowland areas.

In environments other than the tropics, *Nepenthes* need to be grown in glass houses or terrariums, although some growers in subtropical and temperate climates have had success growing highland species outdoors. Once warm conditions are established, these plants are easy enough to look after, will flourish and generally need to be cut back each year.

Nepenthes are vines with jugs or pitchers at the end of each leaf. The vines reach high into the jungle canopy from a forest floor of leached soil. The pitchers are connected to the leaves by a tendril or petiole. The tendril is short on the first few leaves and becomes longer as more leaves appear and the vine grows. The tendril coils up to four times around other branches as the vine requires support to reach the jungle canopy. Not only does the tendril change by coiling and growing longer as the plant grows taller, the pitchers change too. There are two main types of pitcher produced, a lower one that sits on the ground and an upper one that dangles in mid air on the end of the tendril. Depending on the growing conditions, plants will produce a less common intermediate pitcher, longer and thinner than the lower pitcher. Lower pitchers are the first that appear at the base of the plant. They are squat and rounded, their prey crawling insects such as ants and cockroaches. They have a broad rim (or peristome), and often a large spur at the hinge between the lid and the peristome, and two 'wings' running down the front of the pitcher. The wings are usually fringed and, like the pitcher, face the centre of the plant. As the *Nepenthes* grow taller the wings become thinner and less fringed from the base of the pitcher. After the first five to ten leaves appear the shape of the pitcher changes, sometimes to an intermediate shape, facing away from the plant at right angles.

As the *Nepenthes* grows it grips branches with its petioles, and the distance between each leaf (the internode) increases. The leaves become narrower and the upper pitchers begin to form. These pitchers are tapered towards the base, hang in mid-air and attract flying insects such as moths and wasps. Upper pitchers have insignificant wings, more like ribs, and are not fringed. The upper pitchers usually have a thin, flat peristome between the ribs, at the front of the pitcher they face away from the plant, and the spur is usually only a thin spike or slight bump. At this stage the *Nepenthes* is 1–4 m (3.3–13.2 ft) tall, depending on the species and provided there is ample light the inflorescence will begin to form. This inflorescence will be either male or female with dozens of flowers and, if female, the resultant seed capsules will later produce ten thousand or more seeds. By this stage the lower pitchers have turned brown and begun to decay and the base of the stem is woody and brown. The plant will now only produce lower pitchers again if the top is cut off. Plants produce upper pitchers sooner under shaded conditions (low on the forest floor) than if the plant is in full sun on a cliff face.

Some plants have been found growing high in the jungle canopy, growing like an epiphyte in the fork of trees with no apparent root system. These plants are either offshoots of plants that have grown up to the canopy and been separated from the rootstock, or have grown from seeds that have wafted up and germinated in the forks of trees. Both upper and lower pitchers have been observed growing as epiphytes, suggesting both these possibilities. In open fields, where there are no other vines or branches to provide support and shade, *Nepenthes* are shorter and stouter with thicker leaves and stems. The coil of the tendril sits like a spring cushioning the plant against movement of the pitcher.

Some species, for example *N. ampullaria*, produce few upper pitchers—whether this is a result of the plant's natural growth cycle or due to the abundance of food at ground level it is difficult to say. Plants in fertilised cultivation tend to produce fewer pitchers, or even none at all when over fertilised.

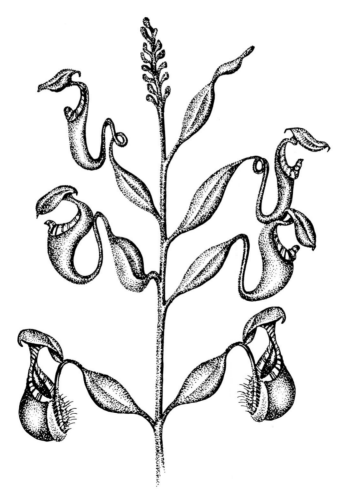

Nepenthes rafflesiana

Nepenthes alata

Common Name: Winged Nepenthes.
Size: Up to 4 m (14 ft) tall.
Climate: Tropical, humid. Both highland and lowland species have been found.
Distribution: The Philippines, Malaysia, Sumatra.
Flowering Time: Spring to summer.
Colour: Variable. Often green in the shade and pinkish in full sunlight. Pitchers are light green with pink to red spots, the rim often red to dark red; leaves dark green with light green underside.
Description: The 4-8 mm (0.16-0.32 in) thick stem can be prostrate or climb to 4 m (14 ft). Leaves are petiolate, 13 cm (5 in) at the base of the plant and 15-25 cm (6-10 in) on the climbing stems. The pitchers, usually cylindrical, are bulbous at the base. The lower pitchers are up to 10 cm (4 in) high and 2.5 cm (1 in) wide with two fringed wings. Upper pitchers are up to 25 cm (10 in) long and 6 cm (2.5 in) wide, lacking wings but having two prominent ribs. There is a spur at the hinge of the lid. Inflorescence up to 60 cm (24 in), with 1 or 2 flowers on each pedicel, and seeds are 1 cm (0.4 in) long.

N. alata is one of the hardiest *Nepenthes*, existing in both high and lowlands, in open fields and in grassy forests and growing alongside *N. merrilliana* and *N. ventricosa*, both of which it hybridizes with.

Nepenthes albo-marginata

Nepenthes alata

Nepenthes albo-marginata

Common Name: Monkey's Rice Pot.
Size: Up to 2 m (6.6 ft) tall.
Climate: Tropical, humid, lowland species.
Distribution: Malaysia, Sumatra, Borneo.
Flowering Time: Spring to summer.
Colour: Pitchers either green with red to purple blotches or burgundy with a white collar below the rim (hence the name)—inside, the pitcher is often purple; leaves are green with a reddish brown underside, or dark burgundy to black if the pitchers are dark burgundy; young parts of the plant are covered in short dense brown hairs, as is the petiole.
Description: A slender vine 3-7 mm (0.12-0.28 in) thick, with large leaves (up to 30 cm (12 in)) close to the base, and leaves 10-25 cm (4-10 in) long and 2-3 cm (0.8-1.2 in) broad further up the climbing stem. Lower pitchers are ovate in their lower half and cylindrical in their upper, 12 cm (4.8 in) long and 2-3 cm (0.8-1.2 in) wide. They have two prominent fringed wings along the length of the pitcher and a large spur behind the lid. Further up the climbing stem the upper pitchers are almost totally cylindrical, up to 20 cm (7.5 in) long, 2-4 cm (0.8-1.6 in) wide at the rim, sometimes narrowing to a 1 cm (0.4 in) 'waist'. Sometimes part of the fringed wings of the pitcher remain near the top. Inflorescence is 20-30 cm (8-12 in) long, and the wrinkled seeds are about 2 cm (0.8 in) long.

Nepenthes ampullaria

Common Name: None.
Size: Up to 6 m (20 ft) tall.
Climate: Tropical, humid, lowland species.
Distribution: Borneo, Malaysia, New Guinea, Sumatra, Singapore; in swampy areas.
Flowering Time: Spring to summer.
Colour: Pitchers vary from near white, to cream with pink to red spots, to yellowish green or green with burgundy spots, and there is even an all red form. Leaves green, sometimes red on the underside; flowers green to brownish green, on a red column.
Description: A woody vine, thick at the base (3-4 cm (1.2-1.6 in)) and slender 5-8 mm (0.2-0.3 in) near the top. The stem can be prostrate in the open. Leaves are 12-25 cm (4.8-10 in) long and 5 cm (2 in) wide. Pitchers form from a short petiole around a rosette which develops from the stem; the rosette may have five or more pitchers. Lower pitchers are squat, horizontally flattened, 2-10 cm (0.8-4 in) high and up to 8 cm (3 in) wide, with two densely fringed wings and a large spur behind the lid. The rim is usually narrow. The lid curls and folds backwards when the pitcher matures, exposing the pitcher to rain. (A freshwater crab, *Geosesarma malayanum*, can live in the pitcher.) Upper pitchers are rare. Inflorescence is 20-40 cm (8-16 in), and has up to 10 flowers per inflorescence. The seeds are 10-15 mm (0.4-0.6 in) long. Some naturally occurring hybrids exist, including the particularly attractive *N. ampullaria* x *rafflesiana*, and *N. ampullaria* x *gracilis*. *N. ampullaria* both hybridises with and grows alongside *N. mirabilis* and *N. gracilis*.

Nepenthes bicalcarata

Nepenthes bicalcarata

Common Name: None.
Size: Up to 15 m (50 ft) tall.
Climate: Tropical, humid, lowland species.
Distribution: Borneo.
Flowering Time: Spring to summer.
Colour: Pitchers rust coloured; leaves dark green, light green on the underside.
Description: A climbing stem, 6-16 mm (0.25-0.6 in) thick, with 20-65 cm (8-26 in) long leaves, 6-12 cm (2.4-4.8 in) broad. Lower pitchers are squat (6-10 cm (2.4-4 in)), with two fringed wings divided by a flat area, running the length of the pitcher, while the rest of the pitcher is round. The two curved thorns under the lid explain the name, bicalcarata, meaning two spurs—these spurs appear to have no function. The yellow-green peristome slopes towards the wings in the front and rises quite a distance to almost a right angle at the back, connecting the lid. Upper pitchers are also ribbed and are slightly longer (13 cm (5 in)) and not as broad (3-8 cm (1.2-3 in)). Inflorescence is up to 1 m (3.3 ft) with 4-15 flowers on each pedicel. *N. bicalcarata* are common to the wet, sandy, dense forest where the soil is unusually wet for *Nepenthes* but is ideally suited to *N. bicalcarata* and to the cane palms that it uses as a support to reach the sunlight.

Nepenthes ampullaria

Nepenthes fusca

Common Name: None.
Size: Up to 10 m (33 ft) tall.
Climate: Tropical, humid, highlands.
Distribution: Borneo.
Flowering Time: Spring to summer.
Colour: Lower pitchers are purple to black with green spots (hence the name *fusca* meaning blackish); upper pitchers light green with burgundy spots; leaves dark green, light green on the underside.
Description: Stem about 7 mm (0.3 in) thick, slightly flattened in the lower part. Leaves petiolate and leathery, 12-15 cm (4.8-6 in) long. Lower pitchers about 15 cm (6 in) long, 4 cm (1.6 in) wide and shaped like a test tube. Two prominent wings run the length of the pitcher, and the burgundy-brown peristome curves upward at the back until it is nearly upright. Upper pitchers are a similar size but flare out to a funnel shape, the wings disappear to thin red lines, and the peristome is round and upright, supporting a lid that often grows straight up. Inflorescence is short (15 cm (6 in)), pedicel two-flowered.

N. fusca grows in full sunlight between rocks, often in very dry soil—consequently plants are short and robust. *N. fusca* is often found alongside *N. stenophylla* and *N. reinwardtiana*.

Nepenthes burbidgeae

Nepenthes burbidgeae

Common Name: Painted Pitcher Plant.
Size: 12-15 m high (40-50 ft) tall.
Climate: Tropical, humid, high altitude.
Distribution: Mt Kinabalu, Borneo.
Flowering Time: Spring to summer.
Colour: Lower pitchers are ivory or cream with burgundy blotches; upper pitchers light green with purple or burgundy blotches; leaves purple green, light green on the underside.
Description: A strong 12-15 mm (0.5-0.6 in) thick stem, with leaves 20-35 cm (8-14 in) long and 6-8 cm (2.4-3 in) broad, with a winged petiole and curved tendrils. Lower pitchers are ovate with two fringed wings, 6-15 cm (2.4-6 in) long and 3-12 cm (1.2-4.8 in) wide. The striped burgundy rim is round and elevated towards the lid, behind which is a small spur. Upper pitchers are slightly smaller and more cylindrical, and wings are replaced by prominent ribs. Inflorescence is 25-40 cm (10-16 in) long with two flowers on each pedicel.

Nepenthes burbidgeae

Nepenthes fusca

Nepenthes gymnamphora

Nepenthes gracillima

Common Name: None.
Size: Up to 5 m (16 ft).
Climate: Tropical.
Distribution: Malaysia.
Flowering Time: Spring to summer.
Colour: Leaves green; pitchers splotched purple/black, or (sometimes) totally black, contrasting strongly with the cream to pale green interior.
Description: The long narrow leaves of this *Nepenthes* are 5–14 cm (2–5.5 in) long and up to 4 cm (1.6 in) wide. Lower pitchers are up to 14 cm (5.6 in) long and 3 cm (1.2 in) wide, are cylindrical and sometimes slightly bulbous in their lower third. Upper pitchers grow to 23 cm (9 in) in length and 3 cm (1.2 in) in width. Inflorescence is up to 18 cm (7 in) long; pedicel one-flowered.

Nepenthes gracillima

Nepenthes gymnamphora

Common Name: None.
Size: Up to 15 m (50 ft) or (rarely) up to 40 m (130 ft).
Climate: Tropical, humid, highland species.
Distribution: Borneo, Java, Sumatra.
Flowering Time: Summer.
Colour: Leaves green to reddish brown; pitchers green with purple to red blotches, sometimes almost totally purple/red.
Description: Leaves are up to 35 cm (14 in) long and 6 cm (2.5 in) wide, and the lower stem is usually woody. Lower pitchers are bulbous in the lower half, conical in the upper, 8–12 cm (3–4.75 in) high and 3–4 cm (1.5–1.75 in) wide with two fringed wings and a flattened peristome. Upper pitchers are slightly rounded in their lower sections and cylindrical in the upper, and up to 18 cm (7 in) long and 4 cm (1.75 in) wide with two prominent ribs. Inflorescence is one- to two-flowered at the top of the plant and up to 30 cm (12 in) long.

Nepenthes khasiana

Common Name: None.
Size: Up to 60 cm (24 in) in the wild, but in cultivation it grows taller.
Climate: Subtropical, humid and subhumid, highland species.
Distribution: Khasi Mountains of Assam, India below 1000 m (3200 ft).
Flowering Time: Summer.
Colour: Leaves green; pitchers green, mottled red.
Description: One of the few *Nepenthes* species that is able to withstand cooler day and night conditions, this is a short stocky plant in its natural habitat. Pitchers are tube-like with a bulge in the lower part. Fringed wings appear on lower pitchers with a faint rib on upper pitchers. Lower pitchers are 10–18 cm (4–7 in) long and 4–8 cm (1.6–3.2 in) in diameter. The lid is oval and connected to a wide rim, with a spur where the lid connects to the rim. The raceme is 25–45 cm (10–18 in) long with fruit up to 2.5 cm (1 in) long.

Nepenthes khasiana

Nepenthes leptochila

Nepenthes leptochila

Common Name: None.
Size: Up to 15 m (45 ft) tall.
Climate: Humid, tropical, highland species.
Distribution: Borneo.
Flowering Time: Spring to summer.
Colour: Leaves green; pitchers usually red, or light green in shade.
Description: Climbing stem up to 15 m (45 ft) tall, leaves 15–25 cm (6–10 in) long, 2.5–6 cm (1–2.5 in) wide. Lower pitchers 8 cm (3 in) long and 3 cm (1.25 in) wide with two fringed wings. Upper pitchers are 9 cm (3.5 in) high, 2.5 cm (1 in) wide, slightly bulbous in the lower third and cylindrical in the upper, with two prominent ribs.

Nepenthes leptochila

Nepenthes lowii

Common Name: None.
Size: Up to 8 m (26 ft) tall.
Climate: Tropical, humid, highland species.
Distribution: Borneo.
Flowering Time: Spring to summer.
Colour: Leaves green; pitchers green to red.
Description: Leaves are 15-30 cm (6-12 in) long, and 6-9 cm (2.5-3.25 in) wide. The hourglass-shaped pitchers are 15-25 cm (6-10 in) long, narrowing to a third of this at the 'waist'. They are dark crimson inside, and the inside of the lid is covered with bristly hairs. A spur connects the lid to the rim. Lower pitchers have two prominent ribs, while upper pitchers have only a faint line of ribs. Inflorescence is 15-25 cm (6-10 in) long, pedical two-flowered.

Nepenthes lowii *Nepenthes lowii*

Nepenthes lowii pitchers: (l to r) intermediate, lower and upper.

Nepenthes macfarlaneii

Nepenthes maxima

Common Name: None.
Size: Up to 3 m (18 ft) tall.
Climate: Tropical, humid, highland species.
Distribution: Borneo, Irian Jaya, Papua New Guinea, Sulawesi, Moluccas.
Flowering Time: Spring to summer.
Colour: Pitchers vary, from yellow green to white with burgundy blotches or purple with white blotches; green leaves.
Description: A varied species that often confuses botanists, hence many synonyms. It has a triangular stem, is often winged, and leaves are petiolate, up to 30 cm (12 in) long and 7 cm (2.75 in) broad, with a curling tendril. Lower pitchers are up to 20 cm (8 in) long and 6 cm (2.5 in) broad; the lower two thirds are ovate, while the upper part is cylindrical, with two fringed wings and a wide round peristome narrowing towards the heart-shaped lid. Upper pitchers are often tubular, up to 30 cm (12 in) long and 8 cm (3 in) wide. Inflorescence up to 20 cm (8 in) with 1 flower, and seeds are 2 cm (0.8 in) long and wrinkled.

N. maxima is a popular species for hybridising, especially with lowland species, as they result in a hardy plant.

Nepenthes maxima

Nepenthes macfarlaneii

Common Name: None.
Size: Up to 4 m (12 ft) high.
Climate: Tropical, humid, lowland species.
Distribution: Malaysia.
Flowering Time: Spring.
Colour: Leaves green; upper pitchers pale yellow/cream; lower pitchers greenish-brown with red blotches.
Description: Often only the rim of the pitcher of *N. macfarlaneii* is visible in the thick sphagnum moss: it is a short plant with lanceolate leaves 18 cm (7 in) long and 6 cm (2.5 in) wide. Both upper and lower pitchers have a thin peristome, flattened in the front with a leaf that clasps around the stem. Lower pitchers are bulbous in the lower half and cylindrical in the upper, have two fringed wings, and are up to 20 cm (8 in) long and 7 cm (2.75 in) wide. Upper pitchers have a narrow base are 17 cm (6.6 in) long and 6 cm (2.5 in) wide, with wings that are only a faint line. Inflorescence is 25 cm (10 in) long and two-flowered.

Nepenthes maxima

Nepenthes mirabilis

Common Name: Tropical Pitcher Plant.
Size: Up to 8 m (24 ft) high.
Climate: Tropical, humid, lowland species.
Distribution: Australia, Indo-China, Indonesia, Malaysia, southern China, Papua New Guinea, Philippines (this is the most wide-ranging species of the genus).
Flowering Time: Spring to Summer.
Colour: Leaves green, pitchers reddish brown in full sunlight. occasionally all red; leaves dark green, light green on the underside.
Description: This species has many forms, varying considerably in appearance and size. Leaves are petiolate and finely fringed, with a winged petiole. Lower pitchers are ovate at the base and cylindrical above, with two fringed wings; they can be up to 10 cm (4 in) high and 2 cm (0.8 in) wide. Upper pitchers are cylindrical with prominent ribs, and are up to 16 cm (6.3 in) high and 3 cm (1.25 in) broad. Inside, pitchers are splotched red or pink. The underside of the lid is red, providing a sharp contrast to the green surface, and there is a spur behind the lid where it is attached to the flattened peristome. Inflorescence is up to 40 cm (16 in) long, pedical one-flowered. It is often found in open fields and swamps in full sunlight, where they are quite stocky, seldom growing taller than 75 cm (30 in). On some Pacific Islands, locals chew the seed pods, which taste similar to tobacco.

N. mirabilis is often found alongside *N. ampullaria* and *N. gracilis*, as are hybrids of these species.

Nepenthes pilosa

Nepenthes pilosa

Common Name: None.
Size: 10–15 m (33–49 ft) tall.
Climate: Tropical, humid, highland species.
Distribution: Borneo.
Flowering Time: Spring to summer.
Colour: Leaves green, pitchers reddish brown in full sunlight.
Description: Leaves are 20–30 cm (8–12 in) long and 6–8 cm (2.5–3 in) wide. Lower pitchers are 10 cm (4 in) high, 3–4 cm (1.25–1.5 in) wide and ovate in the lower third. Upper pitchers are up to 18 cm (7 in) high and 8 cm (3 in) wide at the bulbous top, narrowing towards the lower part of the pitcher. Most of the plant is covered in reddish-brown hairs.

Nepenthes mirabilis

Nepenthes mirabilis

Nepenthes rafflesiana

COMMON NAME: None.
SIZE: Up to 4 m (13 ft) or, rarely, to 15 m (45 ft) tall.
CLIMATE: Tropical, humid, lowland species.
DISTRIBUTION: Peninsular Malaysia, Sumatra, Borneo, Singapore, Papua New Guinea, Irian Jaya.
FLOWERING TIME: Spring to summer.
COLOUR: Upper pitchers green with red spots, or white or cream with burgundy blotches in full sunlight; lower pitchers almost burgundy/black with green blotches; leaves dark green, underside light green.
DESCRIPTION: Leaves petiolate, up to 30 cm (12 in) long, 3–10 cm (1.2–4 in) broad; petiole has narrow wings. Lower pitchers 5–25 cm (2–10 in) high and 3–10 cm (1.2–4 in) wide with two very prominent, wide wings; lower part of pitcher broad and rounded, upper part conical; rim widens below the lid. Upper pitchers up to 40 cm (16 in) high and 6 cm (2.5 in) wide with prominent ribs and a peristome that appears pushed up in the front. Inflorescence up to 50 cm (20 in) long, pedicel one-flowered.

N. rafflesiana is often found alongside *N. gracilis* and *N. ampullaria*, as are hybrids of these.

Nepenthes rafflesiana

Nepenthes rajah

Nepenthes rajah

COMMON NAME: None.
SIZE: Up to 2 m (6 ft) tall.
CLIMATE: Tropical, humid, highland species.
DISTRIBUTION: Borneo.
FLOWERING TIME: Spring to summer.
COLOUR: Pitchers burgundy to dark brown; leaves dark green, underside light green.
DESCRIPTION: One of the largest-pitchered and most dramatic of all *Nepenthes*, this is also one of the slowest growing, taking 10 years or more to progress from seed to maturity. It has a coarse stem, leaves up to 50 cm (20 in) long and 15 cm (6 in) wide, very large oval pitchers up to 35 cm (14 in) high and 18 cm (7 in) wide, and two wide fringed wings extending from the base to the rim of the pitcher. The peristome is deep red to burgundy, and under cultivation the inside of the lid and pitcher are yellow green. In cultivation this *Nepenthes* has subtle colours with the pitchers being pinkish and spotted. In older pitchers the lid rises straight up and the peristome slopes down, and developing pitchers, 5–8 cm (2–3 in) long, have wings that are 5 mm (0.2 in) wide. Upper and lower pitchers are similar except upper pitchers are funnel-shaped. Inflorescence up to 80 cm (32 in) long, pedicel two-flowered.

Nepenthes rajah

Nepenthes sanguinea

Common Name: None.
Size: Up to 7 m (21 ft) high.
Climate: Tropical, humid, highland species.
Distribution: Peninsular Malaysia.
Flowering Time: Spring to summer.
Colour: Pitchers green with red or burgundy blotches; leaves green; also a completely burgundy form.
Description: Leathery sessile leaves are up to 20 cm (8 in) long and 5 cm (2 in) wide. Large lower pitchers ovate in the lower part, cylindrical in the upper, with two fringed wings 5-30 cm (2-12 in) high and 3-7 cm (2.25-3 in) broad, mouth nearly horizontal but raised towards the lid. Upper pitchers up to 25 cm (10 in) high and 3-6 cm (2.25-2.5 in) wide, cylindrical and with two prominent ribs. Inflorescence up to 60 cm (24 in) (commonly 40 cm (18 in)) long, pedicel two-flowered. Found at relatively low altitudes and in exposed locations, alongside *N. macfarlanei* and *N. gracillima* and the resultant hybrids.

Nepenthes reinwardtiana

Nepenthes reinwardtiana

Common Name: None.
Size: Up to 20 m (66 ft) high.
Climate: Tropical, humid, lowland species.
Distribution: Borneo, Sumatra, Malaysia.
Flowering Time: Spring to summer.
Colour: Two forms, green and pale red. A distinctive feature is the two highly visible pale yellow spots on the white inside wall of the pitcher facing the opening. Leaves dark green, underside light green.
Description: A triangular, sometimes winged, stem, with leaves up to 20 cm (8 in) long, 1-4 cm (0.4-1.6 in) broad, sessile, decurrent into two long wings. Lower pitchers are 6-15 cm (2.4-6 in) high, and 2.5-5 cm (1-2 in) broad, ovate in the lower part and cylindrical in the upper. They have two wings that are infrequently fringed, mouth rounded, upper pitchers similar to lower ones but more tubular. Inflorescence 15-35 cm (7-14 in) long, pedicel two-flowered.

Nepenthes sanguinea

Nepenthes stenophylla

Common Name: None.
Size: Up to 10 m (30 ft) high.
Climate: Humid, tropical, highland species.
Distribution: Borneo, Malaysia.
Flowering Time: Spring to summer.
Colour: Leaves dark green; pitchers light green to cream with red to brown blotches.
Description: A winged stem climbing up to 10 m (30 ft), with slender leaves 15-23 cm (6-9 in) long and 3-5 cm (1.25-2 in) wide with pubescent edges. The cylindrical pitchers are 10-25 cm (4-10 in) high and 3-5 cm (2.25-2 in) wide with two prominent ribs, fringed at the top. It has a spur at the back where the lid joins the rim. Inflorescence is 20-30 cm (8-12 in) long, pedicel two-flowered.

Nepenthes stenophylla

Nepenthes tentaculata

Common Name: None.
Size: Up to 2 m (6.6 ft) high.
Climate: Tropical, humid, highland species.
Distribution: Borneo, Sulawesi.
Flowering Time: Spring to summer.
Colour: Leaves green to red green; lower pitchers cream to green with burgundy blotches; upper pitchers burgundy to purple.
Description: Leaves are 5-15 cm (2-6 in) long, 1-3 cm (0.4-1.25 in) wide and are amplexicaul (wrap around the stem). Lower pitchers are up to 9 cm (3.5 in) and 4 cm (1.5 in) wide and ovate in the lower third, narrowing towards the mouth with two fringed wings. Upper pitchers are cylindrical, 10-20 cm (4-8 in) high and 2-5 cm (0.5-2 in) wide with two prominent ribs, sometimes fringed. The lid of the pitcher is covered with thick bristles. Inflorescence 5-10 cm (2-4 in) long, pedicel one-flowered.

Nepenthes tentaculata

Nepenthes truncata

Nepenthes veitchii

Nepenthes truncata

Common Name: None.
Size: 4.5 m (15 ft) high.
Climate: Tropical, humid, lowland species.
Distribution: The Philippines.
Flowering Time: Spring.
Colour: Leaves green; pitchers green to purple.
Description: A short stout plant, *N. truncata* is terrestrial or epiphytic. Leaves truncate, 9–60 cm (3.5–24 in) long. Pitchers have a cordate lid with a keel on the underside; the rim is pale green with red/burgundy ridges. Lower pitchers have wide fringed wings, while upper pitchers have no wings and are cylindrical though slightly bulbous in the lower third. Inflorescence is yellowish green and 80–125 cm (32–50 in) long. Seeds are 12–15 mm (0.25–0.50 in) long.

Nepenthes veitchii

Common Name: None.
Size: Up to 1 m (38 in) high.
Climate: Tropical, humid, lowland species.
Distribution: Borneo.
Flowering Time: Spring.
Colour: Leaves green; pitchers pale green to tan with yellow to tan peristome.
Description: This species is usually found as an epiphyte nestled in the fork of a tree. Leaves are 25 cm (10 in) long and 10 cm (4 in) wide, with a quite short tendril, sometimes only two-thirds the length of the leaf. Pitchers bulge towards the middle. The lower pitchers are 28 cm (11 in) long and 10 cm (4 in) wide, curve sideways, and have fringed wings and a very wide (up to 6 cm (2.5 in)) peristome. The smaller upper pitchers are 17 cm (7 in) long and 4 cm (1.6 in) wide with thin, barely visible, wings at their base, becoming more pronounced (up to 5 mm (0.2 in)) at the top. The base of the pitcher is tapered, flaring at the top. There is a spur at the back of the base of the lid, and the leaves, tendril and pitchers are densely pubescent when young and slightly less so when older.

Nepenthes ventricosa

Common Name: None.
Size: 2 m (6.6 ft) in its natural habitat, taller in cultivation.
Climate: Tropical, humid, highland species.
Distribution: The Philippines.
Flowering Time: Summer.
Colour: Leaves green; top half of the pitcher yellow green with red spots, lower half light green to yellow; pitchers have a red rim and a pale green lid on the underside and green with mottled red on top.
Description: This species has long, thin leaves, 30 cm (12 in) long and 2.5 cm (1 in) wide that terminate in a greenish tendril attached to which are hour-glass-shaped pitchers no more than 15 cm (6 in) long, with no wings.

N. ventricosa is also a hardy species, that can be grown outside a glasshouse provided the temperature does not fall below 5°C (41°F). A very attractive plant, it makes an ideal hybrid. A popular naturally-occurring hybrid is *N. x ventrata (N. alata x ventricosa)*.

Nepenthes ventricosa

Nepenthes villosa

Nepenthes villosa

Common Name: None.
Size: Up to 2 m (6 ft) high.
Climate: Tropical, humid, highland species.
Distribution: Mount Kinabalu, Borneo.
Flowering Time: Spring to summer.
Colour: Densely covered in brown hairs; pitchers green, yellow-green or yellow, with burgundy blotches, or almost completely burgundy; leaves light green.
Description: With its sharply toothed peristome this is one of the most striking *Nepenthes*. Stem is climbing or prostrate, the thick leathery leaves petiolate, 10-22 cm (4-8.5 in) long and 5-10 cm (2-4 in) broad. Lower pitchers 5-15 cm (2-6 in) high and wide, almost globose, with two wide wings fringed at the top and a rounded mouth, acute towards the lid. Upper pitchers up to 20 cm (8 in) high and 12 cm (5 in) wide, ovate in lower half and narrower in the top, with two wings. Inflorescence up to 60 cm (24 in) long, pedical one-flowered. Found amongst thick moss forest where the bright pitchers provide a distinctive contrast.

Nepenthes villosa

Nepenthes villosa

Nepenthes x mixta

Nepenthes x rokko

Nepenthes x henryana

Nepenthes formosa

Pinguicula
pin-gwik-ū-la

Pinguicula, of the family Lentibulariaceae, has 54 species in a variety of habitats, including peat bogs, alongside slow-flowing streams and on rocks where water seeps. It is native to humid areas in much of the Northern Hemisphere and in South America. The name derives from the Latin *pinguis*, referring to the greasy leaves (hence the common term Butterworts). *Pinguicula* were first classified by Carl Linnaeus in 1753.

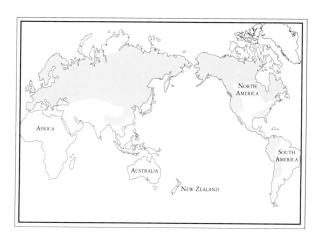

Pinguicula can be divided into three main groups: those that produce winter resting buds (hibernaculia); those that do not produce winter resting buds but have a different leaf from summer to winter (heterophyllous); and, lastly, those that produce the same leaf in both winter and summer (homophyllus). They are relatively easy to grow, especially those that do not form winter resting buds and can be grown in tropical to temperate climates in filtered light. They are perennial, and mostly prefer filtered light. The leaves are usually greenish yellow with hundreds of short fine sessile glands that produce droplets of clear mucilage, creating leaves that gleam silvery in the sunlight. They form flat rosettes up to 20 cm (8 in) in diameter, from the centre of which a flower scape up to 25 cm (10 in) long produces a single flower. The scape is covered with sessile hairs that capture crawling insects, making pollination only possible via flying insects. Flowers appear from spring to summer and have two stamens and five (sometimes heavily veined) petals, two at the top and three at the bottom. These small, thin, papery petals can be pink, white, purple, yellow or violet. Each flower has one large 'spur' at the back and two stamens well hidden inside the corolla. A single plant can produce five or more flower scapes. Roots are thick, white and succulent, prone to breakage and fungal attack—consequently *Pinguicula* should not be potted up during its growing season.

In some species, for example *P. primuliflora*, *P. vulgaris* and *P. grandiflora*, the leaf margins roll up to partially engulf their prey. Sometimes a depression in the leaf appears and aids in engulfing the prey, which may take several days to dissolve.

As to propogation, *Pinguicula* with hibernacula tend to produce gemmae around the base of the resting bud. Gemmae should be removed and potted up before the end of winter, and will produce plants identical to the parent. If left attached, the gemmae can smother and rot off. In their natural habitat gemmae are relocated by rain, snow or animals, thus growing and spreading the species. Some *Pinguicula*, such as *P. primuliflora* and *P. vallisneriifolia*, also produce plantlets on the tips of the leaves that can be removed and potted up. Heterophyllous *Pinguicula*, in contrast, produce progressively shorter leaves during autumn, until the plant resembles a tightly clustered winter rosette.

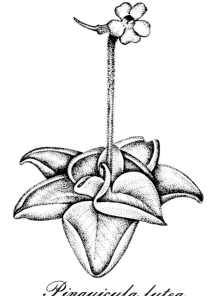

Pinguicula lutea

Pinguicula agnata

Common Name: None.
Size: 10–24 cm (4–9.5 in) in diameter.
Climate: Tropical, humid to subhumid.
Habitat: Damp, mossy limestone cliffs.
Distribution: Mexico.
Flowering Time: Spring.
Colour: Leaves yellowish green or purple; flowers white with pale blue bases and tips; centre of flower yellow to pale green.
Description: Large thick succulent-looking obovate leaves form a rosette 10–24 cm (4–9.5 in) in diameter. The flower is average to large for *Pinguicula* with petals up to 15 mm (0.25 in) long. The scape, spur and five sepals are pale green. The spur is about half the size of the petals, short for a *Pinguicula*. This species does not produce winter resting buds.

Pinguicula alpina

Common Name: Alpine Butterwort, Pale Butterwort.
Size: 5 cm (2 in) in diameter.
Climate: Alpine.
Habitat: Lime-rich soil.
Distribution: Mountainous limestone areas of Europe and Asia.
Flowering Time: Summer.
Colour: Petals white with a centre of yellow-orange; leaves pale green.
Description: Leaves with distinctly curled edges form a star-shaped rosette 5 cm (2 in) in diameter with a scape up to 10 cm (4 in). The plant forms winter resting buds for up to nine months of the year, and grows on north-facing slopes.

Pinguicula agnata

Pinguicula alpina

Pinguicula esseriana

Common Name: None.
Size: 3 cm (1.25 in) in diameter.
Climate: Tropical, humid.
Habitat: Damp sandy soils.
Distribution: Mexico.
Flowering Time: Mid-winter to spring.
Colour: Leaves pale green to yellow; flowers pale pink to purple.
Description: A compact perennial rosette species that has, during winter, 40 or more saucer shaped leaves 7 mm (0.25 in) long and 5 mm (0.2 in) wide. The summer rosette is slightly bigger (up to 60 leaves) and darker than in winter. Each plant can produce up to five scapes 11 cm (4.3 in) long with sparse glandular hairs. Flowers are 2.5 cm (1 in) in diameter.

Pinguicula colimensis

Pinguicula colimensis

Common Name: None.
Size: 10 cm (4 in) in diameter.
Climate: Tropical, humid, with cool winters.
Habitat: Sunny mountain slopes at altitudes of 500 m (1500 ft).
Distribution: Mexico.
Flowering Time: Mid-summer.
Colour: Leaves pinkish green; flowers reddish purple.
Description: Previously known as *P. pulcherrima*, and considered one of the most beautiful of all *Pinguicula*, it is in rosette form, fringed with six or more oval leaves 5 cm (2 in) long and 3 cm (1.25 in) wide, turned up at the sides; winter leaves are thicker than the summer version. The flower is 6 cm (2.25 in) wide on a scape up to 14 cm (5.5 in) high, each plant producing two or more scapes. As the leaves press down on the soil, and the plants have very short roots, it does tend to uproot over time.

Pinguicula esseriana

Pinguicula longifolia

COMMON NAME: None.
SIZE: 10–15 cm (4–6 in) in diameter.
CLIMATE: Temperate, humid to subhumid.
HABITAT: Damp limestone cliffs.
DISTRIBUTION: Temperate Europe.
FLOWERING TIME: Summer.
COLOUR: Leaves pale yellowish green; flowers purple with white throats.
DESCRIPTION: The long leaves have pointed tips and are heterophyllous. The purple or pale lavender flowers are bearded with white and have a long flower spur. *P. longifolia* forms a winter resting bud, or hibernaculum, surrounded by smaller buds, or gemmae, which look like light brown grains of rice poked into the sand.

Pinguicula lusitanica

Pinguicula lusitanica

COMMON NAME: Pale Butterwort.
SIZE: 5 cm (2 in) in diameter.
CLIMATE: Temperate-boreal, subhumid-humid.
HABITAT: Damp mountainous areas, acid bogs.
DISTRIBUTION: France, Portugal, Spain, UK.
FLOWERING TIME: Spring.
COLOUR: Leaves yellowish green; flower pink to red through lilac, or pale yellow with lilac veins.
DESCRIPTION: Ideal for terrariums, this is a rosette species with almost linear leaves up to 2.5 cm (1 in) long and 8 mm (0.3 in) wide with prominent rolled edges. Thin and very pale, the leaves appear almost translucent: thin pinkish veins run through them and in full sunlight they darken until the plant appears greenish brown. Summer and winter leaves are about the same size, and gemmae are produced during winter. Plants may have up to three scapes up to 15 cm (6 in) high, each bearing a single trumpet-shaped flower 6 mm (0.25 in) in diameter.

Pinguicula longifolia

Pinguicula moranensis

Common Name: None.
Size: Up to 24 cm (9.5 in) in diameter.
Climate: Subtropical, subhumid.
Habitat: Damp mossy soil, with filtered light provided by surrounding vegetation.
Distribution: Mexico.
Flowering Time: Spring.
Colour: Leaves green; flowers lavender or pink and white.
Description: Fine broad leaves, rounded at the tip. Although *P. moranensis* does not produce winter resting buds, winter leaves are usually smaller. The five petals, each about 3 cm (1.5 in) across, have definite veining from the tip to the white centre—with increased sunlight this veining is less obvious and the overall effect is deep pink to red. The spur is quite long, 3-4 cm (1.5-2 in), and the scape is 16 cm (6.25 in) long, greenish brown in filtered light and pink in full sunlight.

Pinguicula moranensis

Pinguicula pumila

Pinguicula moranensis

Pinguicula pumila

Common Name: None.
Size: Up to 2 cm (0.75 in) in diameter.
Climate: Humid to subhumid, tropical to subtropical.
Habitat: Wet sandy soils, acidic soils.
Distribution: USA.
Flowering Time: Spring.
Colour: Leaves light green; flowers usually white, or lilac, yellow or pink in warmer areas.
Description: A tiny rosette, with egg-shaped leaves up to 1 cm (0.4 in) long and 7 mm (0.27 in) wide, with rolled edges and pointed tips. Flowers up to 7 mm (0.27 in) in diameter appear on two or more scapes up to 6 cm (2.5 in) high, and leaves sometimes form plantlets that can be removed and planted out.

Pinguicula primuliflora

Common Name: None.
Size: 10-15 cm (4-6 in) in diameter.
Climate: Tropical and subtropical, humid to subhumid.
Habitat: Wet areas, often in sphagnum mounds and damp rocks along banks of streams and swampy open grass fields, and on the edges of pine forests.
Distribution: South east coast of North America.
Flowering Time: Spring.
Colour: Leaves pale green; flower has violet petals and a white centre.
Description: Rectangular leaves, which have rolled edges and are rounded at the tip. This is one of the few *Pinguicula* to produce small plantlets on the tips of older leaves during summer, giving the appearance of baby plants circling their mother. Flower scapes reach 3 cm (1.2 in) high and the flowers are 14 mm (0.6 in) in diameter. *P. primuliflora* does not produce winter resting buds and the numerous brown seeds are less than 1 mm (0.04 in) in diameter.

Pinguicula primuliflora

Pinguicula rotundiflora

Common Name: None.
Size: 32 mm (1.25 in) in diameter.
Climate: Tropical, humid.
Habitat: Damp peat soil in winter, drier soil in summer.
Distribution: Mexico.
Flowering Time: Winter.
Colour: Flowers purple; leaves pale green.
Description: A relatively new species renamed in 1985 by Miloslav Studnicka, *P. rotundiflora* measures 32 mm (1.5 in) across, with rounded spathulate leaves 4 mm (0.16 in) wide, with rolled edges. During winter leaves are narrower and non-carnivorous and the plant is slightly smaller. One or two flower scapes reach 7 cm (2.75 in) high, holding flowers measuring 18 mm (0.7 in) across.

Pinguicula rotundiflora

Pinguicula vulgaris

Common Name: Common Butterwort.
Size: 5–10 cm (2–4 in) diameter.
Climate: Subhumid, temperate.
Habitat: Around bogs, open grass or sphagnum fields or bogs in rocky mountainous areas.
Distribution: North America, Europe.
Flowering Time: Summer.
Colour: Leaves green with yellow tips; flowers violet.
Description: From a distance the plants look like small green stars with yellow tips. They form a perennial winter hibernaculum. The flower looks like a small violet with a three lobed lower lip that is slightly bearded with small white hairs. Flowers grow singly on tall scapes up to 10–15 cm (4–6 in) high. It often grows alongside *Darlingtonia californica*, *Drosera intermedia*, and *P. grandiflora* (with which it hybridises).

Pinguicula zecheri

Pinguicula zecheri

Pinguicula zecheri

Common Name: None.
Size: 14 cm (5.5 in) in diameter in summer, 6 cm (2.5 in) in winter.
Climate: Tropical, humid.
Habitat: Damp places amongst sphagnum moss, at high altitudes.
Flowering Time: Late spring to autumn.
Colour: Leaves light green; flowers purple with a white streak.
Description: A perennial rosette *Pinguicula* with up to 50 winter spathulate leaves, each 2 cm (0.75 in) long and 0.5 cm (0.2 in) wide. Summer leaves number 10–15 and are 7 cm (2.5 in) long and 4 cm (1.6 in) wide. Each plant can produce up to four scapes up to 10 cm (4 in) long covered densely with glandular hairs, with flowers 3.5 cm (1.5 in) in diameter.

Pinguicula vulgaris

Sarracenia
săr-ă-sĕn-ĭ-yă

Sarracenia, of the family Sarraceniaceae, has eight species, all native to the south eastern side of North America. They are found in swamps, bogs, wetlands and on the edge of pine forest. The first species of *Sarracenia* was described by the French botanist Tournefort in 1700, who named the genus after Michael Sarrazin, who had sent him examples of *S. purpurea*.

Six of the eight species are similar in shape, all having upright tubes with an opening at one end. *S. purpurea* and *S. psittacina*, in contrast, lie prostrate. All species, however, have a central wing (or 'ala') that provides support and prevents the pitchers from blowing over.

Most species can be grown in tropical to temperate climates, and *S. purpurea* can be grown in areas bordering on the boreal. All species can be grown outside in full sunlight provided they have access to water, especially during summer, and are protected from severe frosts. Most also benefit from burning off during summer. *Sarracenia* hybridise freely.

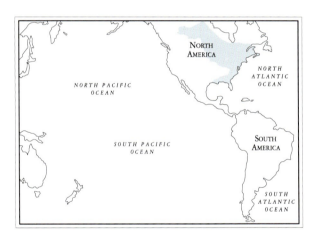

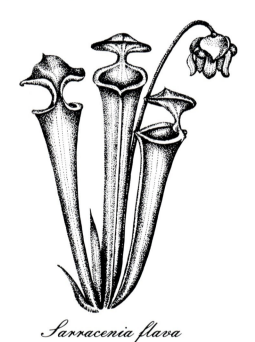

Sarracenia flava

Sarracenia alata

Common Name: Flycatcher, Old Pale Pitcher Plant.
Size: Up to 70 cm (28 in) tall.
Climate: Temperate, humid.
Distribution: Alabama, Texas.
Flowering Time: Early to mid spring.
Colour: Trumpet-shaped pitchers are light green to yellow, some with red veining; flowers white to yellow. Some varieties are red on the lid and a smaller number are short and hairy.
Description: Similar to *S. flava*, without the waisted throat on the pitcher and with a more prominent wing. The tall pitchers widen very gradually to a 5 cm (2 in) mouth; the inside of the lid is often red. The flowers are on scapes up to 75 cm (30 in) high. During winter, older plants may produce phyllodia. Pitchers have fine red veining inside the column.

The moth Exyra can be a problem, and can cause a whole field to appear as though the pitchers are slashed.

Sarracenia alata

Sarracenia leucophylla

Common Name: White Trumpet.
Size: Up to 60 cm (24 in) high.
Climate: Warm, subhumid.
Distribution: South east coast of North America.
Flowering Time: Early to mid spring, or a month earlier under cultivation.
Colour: Pitchers light green graduating to darker green with small white spots, bordered by red veins; new pitchers reddish brown, losing colour as winter approaches; top rim sometimes solid white with a fine red rim; lid near white with red veins during summer and towards the end of the growing season, or white with green veins in poor light.
Description: Pitchers grow to a height of 90 cm (36 in) under ideal conditions and, as one of the tallest *Sarracenia*, the only tall pitcher plant with fenestrated leaves, it is easily distinguishable. The rim of the pitcher bends downwards like the spout of a jug, and it has a wing that runs its full length, less than 1 cm (0.4 in) at its widest point (halfway up the pitcher) narrowing almost to extinction. Pitchers do not form properly in dry soil, and look like long blades of grass, brown at the top. This species adapts to varied conditions, increasing the length of the roots or letting some rot off to meet the appropriate water level. Flowers are on long slender scapes about 10 cm (4 in) above the pitcher. When pitchers reach a height of 60 cm (24 in) the flower scape stops growing and begins to thicken. Flowers are noddling, with a vivid splash of burgundy on the petals, sepal, style, bracts and top 5 cm (2 in) of the scape. In an open field the effect is of a layer of deep red hanging over a white haze.
Varieties and forms: *S. leucophylla* var *alba* (Southern Alabama). There is also a semi-albino form that appears bright white with light green veining.

Sarracenia minor

Sarracenia leucophylla

Sarracenia minor

Common Name: Hooded pitcher plant.
Size: Up to 80 cm (32 in), or (commonly) to 60 cm (24 in).
Climate: Temperate, humid.
Distribution: North and South Carolina, Georgia, Florida.
Flowering Time: Spring.
Colour: Pitchers light green, upper part red in full sunlight, with fenestrations; flowers yellow.
Description: The red-veined lid of the pitcher is a hood that curls over and obscures the mouth of the pitcher preventing rain from entering. There are fenestrations on the upper half of the back of the pitcher that let light inside. Flowers are small, on scapes up to 60 cm (24 in) long. This is one of the most interesting and attractive *Sarracenia* to hybridise. In its natural habitat (often floating on sphagnum islands) natural hybrids occur with *S. purpurea*, *S. psittacina* and *S. leucophylla*.

Sarracenia oreophila

Sarracenia psittacina

Common Name: Parrot Pitcher Plant, Lobster Pot.
Size: Up to 50 cm (20 in) in diameter.
Climate: Temperate, humid.
Distribution: Georgia, Florida, Alabama, Louisiana.
Flowering Time: Mid-spring.
Colour: Pitchers green, or dark red to burgundy with more prominent fenestrations in full sunlight; flowers red.
Description: Pitchers initially grow upwards but as the hood thickens the pitchers fall back to lie horizontally, resulting in a rosette shaped plant, each pitcher opening facing the centre. Pitchers are up to 25 cm (10 in) long with a prominent wing. The hood is large and almost globose over a very small mouth. The dark, small flowers are on 25 cm (10 in) scapes.

This species prefers very wet soil and often floats in bogs with sphagnum moss. It can survive total immersion in floods, and some botanists suggest its trapping abilities are improved by opportunities to trap small swimming prey. *S. psittacina* responds extremely well to fertiliser. When the soil is drier hoods become smaller and thinner and, in very dry soil, non-existent, and the plant becomes non-carnivorous.

S. psittacina is often found with *S. leucophylla*, *S. purpurea* and *S. minor*, all of which it naturally hybridizes with.

Sarracenia oreophila

Common Name: Frog Bonnets, Bugle Grass, Green Pitcher Plant.
Size: Up to 60 cm (24 in).
Climate: Temperate, humid.
Distribution: North Carolina, Georgia.
Flowering Time: Spring.
Colour: Pitchers green with red veins in the shade, or entirely red in full sunlight, especially at the onset of winter; flowers pale yellow.
Description: In cooler climates the pitchers are often hairy, pitchers have an ala running their length, widening at the middle to half the width of the pitcher. Pitchers stand up to 60 cm (24 in) with prominent burgundy veins running from the lid halfway down the plant. The heart-shaped hood with its prominent rib overhangs the mouth of the pitcher. In winter pitchers wither early and the plant produces phyllodia. Flowers are on scapes to 63 cm (25 in) tall. Unusually, only one to two of the sweet smelling noddling flowers are produced per plant. This species was placed on the US endangered species list in 1979.

S. oreophila is found in mountainous regions, growing amongst mosses and ferns in a sand and peat soil. Few other *Sarracenia* species grow near it, preventing natural hybridisation.

Sarracenia psittacina

Sarracenia purpurea ssp *venosa*

Sarracenia purpurea ssp *heterophylla*

Sarracenia purpurea

Common Name: Sidesaddle Plant, Huntsman's Cap.
Size: About 20-30 cm (8-12 in), or, in cultivation, up to 90 cm (36 in) in diameter.
Climate: Temperate, humid.
Distribution: Canada west to the Rocky Mountains (north eastern British Columbia), southwards along the east coast of USA to the Gulf of Mexico.
Flowering Time: Late spring to early summer in the south and mid summer in the north.
Colour: Pitchers green to yellow green, or red, often dark red with the onset of winter; flowers green, pink or burgundy.
Description: Pitchers are prostrate, sometimes hairy, up to 45 cm (18 in) long with a very prominent wing running the full length of the front of the pitcher. They are swollen in the middle and can be up to 10 cm (4 in) wide with a ruffled or unruffled hood with or without wings and form from a rhyzome in a rosette cluster. Flowers grow on a scape up to 45 cm (18 in) high.

Often found floating on sphagnum moss in bogs along with *Drosera anglica* and *D. rotundifolia*, it can be submerged in the sphagnum with only a small opening exposed.

So prolific is *S. purpurea* that a single plant introduced to a 6 ha (15 acre) floating sphagnum island in Ohio, USA in the 1910s produced 157000 plants by the 1960s. It is also one of the hardiest *Sarracenia*, able to cope with both the tropics and snow. As with *S. psittacina*, leaf cuttings are an effective method of propagation, and a leaf may be harvested from an adult plant every month. Many natural hybrids occur, species found alongside *S. purpurea* are *S. leucophylla*, *S. alata* and *S. minor*.

Sarracenia purpurea sub species:
S. purpupea ssp *purpurea heterophylla*—pitchers green only.
S. purpurea ssp *venosa*—pitchers short and fat, green to burgundy with a frilled hood.
S. purpurea purpurea—pitchers long and narrow, green to burgundy, unfrilled hood.

Sarracenia purpurea ssp *venosa*

Sarracenia rubra

Common Name: Sweet Pitcher Plant.
Size: Up to 50 cm (20 in).
Climate: Temperate, humid.
Distribution: North and South Carolina, Florida, Alabama.
Flowering Time: Early to mid-spring.
Colour: Pitchers have red veins on a green background, but sometimes the veins are so profuse the pitcher looks brown; flowers red.
Description: There are a number of subspecies existing including *S. rubra* ssp *rubra*, *S. rubra* ssp *alabamensis*, *S. rubra* ssp *gulfensis*, *S. rubra* ssp *jonesii*, *S. rubra* ssp *wherryi*. All varieties have hairy pitchers standing up to 50 cm (20 in) tall. They are quite thin with an ala running most of the length of the pitcher. Red veining is prominent on both the inside and outside of the pitcher, while the rim is usually pale green, and dips down in front. The numerous small, sweet-smelling flowers are bright red on scapes up to 55 cm (22 in) long.

Sarracenia rubra ssp *alabamensis*

Sarracenia rubra ssp *gulfensis*

Sarracenia rubra ssp *wherryi*

Triphyophyllum
trī-fō-fī-lim

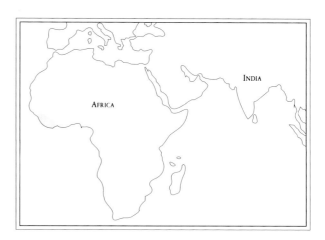

Triphyophyllum, of the family Dioncophyllaceae, is a monotypic genus native to secondary rainforest areas of Liberia, Sierra Leone and the Ivory Coast in West Africa. Originally named *Ouratea glomerata*, it was changed by Airy-Shaw in 1952 to *Triphyophyllum* from the Greek indicating that it has three different types of leaves.

The areas where *Triphyophyllum* grows have an average temperature of 25–30°C (77–86°F) with high humidity, and have a two-season cycle—wet and dry—each lasting six months. During the dry season the leached soil can become rock hard, so to compensate *Triphyophyllum* receives its moisture via a long root that penetrates well below the surface, similar to *Drosophyllum*. In the wet season, of course, this is not a problem. *Triphyophyllum* also grows in filtered light.

Triphyophyllum peltatum

Triphyophyllum peltatum

Common Name: None.
Size: Up to 50 m (160 ft).
Climate: Tropical, humid.
Distribution: Liberia, Sierra Leone and the Ivory Coast in West Africa.
Flowering Time: Spring.
Colour: Leaves green with reddish hairs; flowers white.
Description: *T. peltatum* is a vine which, like *Nepenthes*, uses the ends of its leaves to grip trees and branches to climb to the top of the jungle canopy. It is distinctive in having three different types of leaf at various stages of its growth. The growth cycle of *T. peltatum* begins with the emergence of green lanceolate leaves, 4 cm (1.5 in) wide and 15 cm (6 in) long. Leaves continue to emerge from the tip of a stem that begins to thicken to about 1 cm (0.25 in) as older leaves die off. It reaches a height of about 10 cm (4 in) after about 20 leaves have developed. At this point, a second type of leaf unfurls. It has a midrib, is without a blade, and is twice the length of the earlier lanceolate leaves. This midrib has glandular hairs, each with a red tip, resulting in an overall pink appearance. These are the carnivorous leaves and, like those of *Drosophyllum* and *Pinguicula*, they drown the prey (mainly beetles and ants) as the insects brush past the sensile glands. Some leaves will now be just a midrib with tentacles, others will be half midrib half lanceolate leaf. The plant then alternates between a fully lanceolate leaf and a midrib with tentacles for the next 5-10 leaves till the plant reaches a height of 50 cm (20 in). At this height the third leaf form appears. This is not glandular, but has a leaf-blade terminating in two hooks. It is the double-hooked leaf that clings onto branches, enabling the vine to climb, reach the jungle canopy, and flower.

As *T. peltatum* grows taller (as high as 50 cm (160 ft)), the base becomes thicker (8-10 cm (3-4 in)) and woody. During spring, as the plant reaches the canopy and sunlight, short branches (4 cm (1.5 in)) form with a few white flowers with waxy petals on the ends. The flowers have ten stamens, five branching styles and produce large saucer-shaped bright red seeds 6-7 cm (2.5-2.75 in) in diameter. These seeds are very light and glide to the forest floor to germinate during the wet season within a few weeks.

Interestingly, when cut back to a height of 50 cm (20 in), *T. peltatum* reverts to producing just carnivorous leaves, and may even produce them at ground level.

Triphyophyllum peltatum

Triphyophyllum peltatum

Utricularia

yū-trik-yū-'lā-rē-ă

Utricularia, of the family Lentibulariaceae, has the largest number of species and the smallest traps of all carnivorous plants. The 214 species of this genus are found in almost every country in the world. Carl Linnaeus classified *Utricularia* in 1753. The name derives from the Latin for bagpipe, referring to the little bags that were thought to help the plant stay afloat, hence the common name of Bladderwort. During the 1870s, however, it was realised the bladder traps prey. Until recently *Polypompholyx* (Pink Petticoats) and *Biovularia* were classified as separate genera to *Utricularia*, but both are now classed as species within the *Utricularia* genus. Over time the number of *Utricularia* species is expected to change as more species are discovered in this highly variable family.

Some *Utricularia* are perennials and others annuals. Some are affixed aquatics with the plant floating near the water surface but attached to the soil below—others are free floating. Most (80 per cent) grow on land with their tiny pouches hidden below the surface. Remarkably, some species even grow inside other carnivorous plants. What all *Utricularia* have in common, however, is a small upright flower.

The flower (up to 5 cm (2 in)) across is either pink, yellow, white, blue or red, with one to thirty or so flowers appearing on each scape. Many species can produce two different types of flowers, one chasmogamous (open, that insects can pollinate) and one cleistogamous (insects cannot enter and it is believed it cannot be cross-pollinated). Interestingly, both can produce hundreds of viable seeds, some producing thousands. Some flowers have a long downward-pointing spur, others have long thin erect antennae. What produces this variation in flower shape is unknown, but believed to relate to environmental changes such as decreases in light or changes in water level.

Utricularia gibba

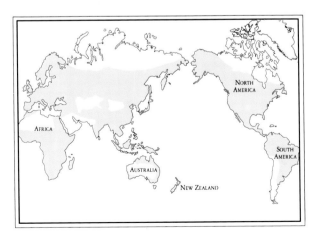

Another feature common to all *Utricularia* is an arrangement of tiny bladders the size of pin heads (or less), which have the most complicated trap of all carnivorous plants. Each trap is water filled, and may be oval, round or slightly tubular. Traps are connected to the plant via stalks, the stolon or leaves; each has a door, with thin, hair-like appendages that direct the prey to the opening. The prey of *Utricularia* is generally small creatures like fish fry and mosquito larvae, although some Australian species are believed to be proficient at catching the small tadpoles of cane toads, a toad introduced to Australia and now in plague proportions. *Utricularia* is one of the few natural controls considered for controlling the toad plague.

Species of *Utricularia* are defined by, among other things, the shape and arrangement of the bladder on each shoot, and the bladder door and its position and the appendages. Other features providing taxonomical information are the leaves and the inflorescence, the flowers and seeds.

Leaves appear in a variety of forms, producing rosettes from a central point, usually the base of the flower scape. Single leaves can develop at regular intervals along a stolon, or clumps of leaves can develop randomly along stolons. Leaf shapes vary from oval to linear, from short leaves like fine blades of grass to species with no obvious leaves visible at flowering time. Many plants would be overlooked if it were not for the flower stem rising from the soil or water.

Many aquatic species develop a food storage unit called a turion, that grows along the stem. During summer the stems are partly filled with air to keep the plant afloat. As winter approaches and the leaves die, the stem floods and sinks to the bottom of the pond or dam, taking the turion with it. From this turion, the plant will emerge and float to the water surface as spring approaches. Terrestrial species can produce perennating structures, usually tubers—in response to decreased water levels—to aid survival during dry periods: they can be as small as grains of rice.

Utricularia alpina

Utricularia alpina

Common Name: Alpine Bladderwort.
Size: Leaves up to 15 cm (6 in) long.
Climate: Tropical, humid.
Habitat: Rainforests; an epiphytic species that grows on moss on trees, usually at high altitudes.
Distribution: Central and South America, the West Indies.
Flowering Time: Late winter.
Colour: Leaves dark green; flowers white and yellow.
Description: A perennial terrestrial or epiphytic species with one of the largest flowers of all *Utricularia*, measuring up to 4 cm (1.5 in) across. It also has elongated, slightly oval-shaped leaves 8-15 cm (3-6 in) under cultivation and up to 5 cm (2 in) in the wild, growing from a rhizome that clings to the wet bark or moss of a tree. Tubers form from the stolon; they are usually white but may turn green when exposed to light. It has a long scape, up to 30 cm (12 in) bearing one to four flowers that hang down from the scape.

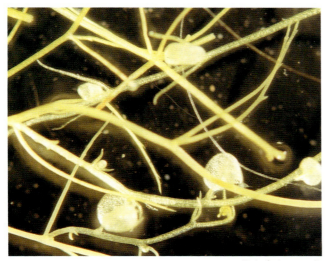

Utricularia australis

Utricularia australis

Common Name: Yellow Bladderwort.
Size: Stems 30-90 cm (12-36 in).
Climate: Tropical to temperate, humid to subhumid.
Habitat: In dams, ponds and slow-flowing creeks.
Distribution: Australia, Japan, New Zealand and Africa and Europe.
Flowering Time: Spring to late summer.
Colour: Flowers sulphur yellow.
Description: This free floating aquatic species floats on the surface and has submerged leaves with traps up to 4 mm (0.16 in) across. A long central stem with many branches is 30-90 cm (12-36 in) long. Two or more flowers 18 mm (0.5 in) across appear on the scape up to 16 cm (6.25 in) long. When the water temperature is too cold during winter, resting buds form at the end of the long stem and sink to the bottom of the dam or pond. This adaptability, and its distribution, makes this species one of the easiest to grow.

Utricularia chrysantha

Utricularia chrysantha

Common Name: Sun Bladderwort.
Size: Leaves up to 5 mm (0.2 in) long.
Climate: Tropical, humid.
Habitat: Swampy areas.
Distribution: Northern Australia, Northern Papua New Guinea.
Flowering Time: Autumn to winter.
Colour: Green leaves; pale to bright yellow flower.
Description: An annual previously known as *U. flava* and *U. barbata, U. chrysantha* has linear leaves up to 5 mm (0.2 in) long and 0.7 mm (0.04 in) wide. The traps are up to 0.6 mm (0.02 in) long, and the scape is up to 60 cm (24 in) long, with up to 20 four-lobed flowers 1.5 cm (0.5 in) wide. Leaves are usually absent while the plants are in flower.

Utricularia cornuta

Utricularia cornuta

Common Name: None.
Size: Leaves up to 4 cm (1.6 in) long.
Habitat: Swamps and pools.
Climate: Tropical-subtropical. humid-subhumid.
Distribution: Bahamas, Cuba, North America.
Flowering Time: Spring to early summer in southern areas. summer in northern areas.
Colour: Leaves green; flowers yellow.
Description: A terrestrial species with rounded leaves up to 4 cm (1.6 in) long. The traps are up to 8 mm (0.3 in) long, and it has a single scape up to 40 cm (16 in) high with up to six flowers 2 cm (0.75 in) in diameter. grouped together at the apex.

Utricularia dichotoma

Common Name: Fairy Aprons.
Size: Leaves 2–25 mm (0.08–1 in) long.
Climate: Tropical to temperate, humid to subhumid.
Habitat: Boggy soil and pools.
Distribution: Eastern states of Australia and Western Australia.
Flowering Time: Summer.
Colour: Leaves green; flower dark purple to pale pink with a white or yellow centre.
Description: A terrestrial species extremely variable in size, with thin linear green leaves. The flower scape is 2–30 cm (0.75–12 in) long with flowers up to 2 cm (0.75 in) wide that usually appear in pairs, opposite one another. The tiny traps lie on the surface, just below a thin water layer.

Utricularia dichotoma

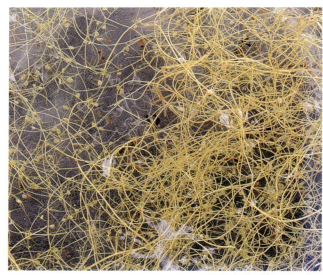

Utricularia gibba ssp *exoleta*

Utricularia inflata

Common Name: None.
Size: Leaves 18 cm (7 in) long.
Climate: Tropical-sub tropical, humid-subhumid.
Habitat: Swamps, lakes and pools.
Distribution: North America.
Flowering Time: Winter to early summer.
Colour: Leaves green; flowers yellow with a darker centre.
Description: A large perrenial aquatic species, this plant is named for the air sacs that allow it to float. It has five or even ten of these sacs, each up to 10 cm (4 in) long, which radiate from the base of the stem and support it. *U. inflata* also has up to ten filiform leaves 18 cm (7 in) long, and a scape up to 25 cm (10 in) long with up to 17 flowers, each about 3 cm (1.5 in) in diameter. Traps are ovoid and up to 3 mm (0.12 in) wide. If the water level in which this species lives is reduced, it produces tubers to survive the dry spell.

U. inflata was previously known as *U. ceratophylla*.

Utricularia gibba

Common Name: None.
Size: Leaves 3–10 mm (0.12–0.4 in) long.
Climate: Tropical, humid.
Habitat: Lakes, ponds and sphagnum hollows.
Distribution: North and South America, Africa, Asia and Australia.
Flowering Time: Mid-summer.
Colour: Flowers sulphur yellow with a red tinge in the centre.
Description: *U. gibba* is one of the many aquatic *Utricularia* that can be grown floating on a fish tank, in a glass on a window sill, or in a sphagnum swamp. An aquatic and sometimes partly terrestrial species, it has a flower scape 5–8 cm (2–3 in) long, bearing two or more flowers, each up to 1 cm (0.4 in) across. The corolla rounds out, and below it the spur points out. As with most aquatic species this plant forms a thick matt of tangled leaves and bladders that are 2–3 cm (0.75–1.2 in) across.

Utricularia inflata

Utricularia menziesii

Common Name: Redcoat.
Size: Leaves 2–5 cm (0.8–2 in) long.
Climate: Temperate, dry summer and wet winter.
Habitat: Damp areas, often near rivulets, in wet coarse clay.
Distribution: South west Western Australia.
Colour: Leaves green; flowers orange to burgundy with a yellow centre.
Description: The short green linear leaves of this terrestrial species look like very short blades of grass. Out of this 'grass' rises a single bright orange to burgundy flower up to 2 cm (0.8 in) wide, on a burgundy scape 2–5 cm (0.8–2 in) tall.

An interesting feature of *U. menziesii*'s life cycle is that during the hot dry summer the plant, including the leaves, dries up and forms a small tuber similar in size and shape to a grain of rice. As rainfall increases and winter begins, leaves emerge in clusters from the tuber, followed by a flower.

Utricularia livida

Utricularia livida

Common Name: None.
Size: Leaves up to 7 cm (2.75 in) long.
Climate: Tropical, subhumid to humid.
Habitat: Wet bogs and seepage areas at altitudes of 900–2830 m (2900–9200 ft).
Distribution: Ethiopa to South Africa, Madagascar, Mexico.
Flowering Time: Throughout the year while the soil is wet.
Colour: Leaves green; flower violet (usually), yellow, white, or lilac and white.
Description: A terrestrial species with wedge-shaped to obovate leaves 6 mm (0.25 in) long are attached to the rhyzomes, as are the stolons. Stolons help the plant survive the dry summer, and many plants appear each year from the numerous seeds. Each plant produces one very long (up to 80 cm (32 in)) scape, which can produce up to 50 flowers 1.5 cm (0.6 in) in diameter.

The most widely dispersed of all *Utricularia*, *U. livida* has a highly variable flower, resulting in great confusion over its identity, with over 30 synonyms in existence.

Utricularia menziesii

Utricularia multifida

Common Name: Pink Petticoats.
Size: Leaves 1-3 cm (0.4-1.2 in) long.
Climate: Subtropical, subhumid.
Habitat: Moist hollows and areas where seepage occurs.
Distribution: South west Western Australia.
Flowering Time: Spring to summer.
Colour: Leaves green; flowers pink with a yellow centre.
Description: Originally classified as a separate genus (*Polypompholyx*), because of the flower arrangement with its four-lobed calyx, and because of the variation in its bladder shape, it was reclassified as part of the *Utricularia* genus in 1986. It is a terrestrial species, best treated as an annual. Its flowers, of which there are one or more, are three-lobed and 18 mm (0.7 in) wide, and appear on a scape 10-30 cm (4-12 in) long. The narrowly spathulate leaves form a basal rosette and usually disappear before the flower emerges. The bladder trap is up to 2 mm (0.08 in) across. It often grows in the same area as *Byblis gigantea*.

Utricularia multifida

Utricularia pubescens

Common Name: None.
Size: Leaves 5-7 mm (0.2-0.3 in) across.
Climate: Tropical, humid.
Habitat: Wet peaty sandy soil in sphagnum hollows.
Distribution: India, tropical Africa, South America.
Flowering Time: Summer.
Colour: Leaves green; flowers lilac to white through pale blue.
Description: A terrestrial species that looks like a small water lily, its small, green, rounded leaves lie flat on the soil. Two or more large flowers 16 mm (0.6 in) wide are attached to a scape 35 cm (14 in) long.

Utricularia pubescens

Utricularia tricolor

Common Name: None.
Size: Leaves up to 3 cm (1.5 in) in diameter.
Climate: Tropical, humid.
Habitat: Wet savannah.
Distribution: South America.
Flowering Time: All year except winter.
Colour: Green leaves; flower violet to lilac with a yellow centre bordered in white.
Description: A perennial plant with up to three leaves 2–3 cm (1–1.5 in) in diameter on 2 cm (0.75 in) stems attached to the rhyzome. Traps are 2 mm (0.08 in) long, and the scape is up to 30 cm (12 in) long with one to four flowers each 6 mm (0.25 in) in diameter.

U. tricolor was previously known variously as *U. fontana*, *U. gomezii*, *U. montantha*, *U. globularifolia*, *U. fusiformis* and *U. rotundifolia*.

Utricularia tricolor

Utricularia tricolor

Utricularia sandersoni

Common Name: Rabbit ears.
Size: Leaves 8–15 mm (0.3–0.6 in).
Climate: Tropical to subtropical; humid to subhumid.
Habitat: Boggy soil with reeds, in filtered light, and on moss covered rocks. At altitudes of up to 1200 m (3900 ft).
Distribution: South Africa.
Flowering Time: All year.
Colour: Leaves light green to pale yellow; flower white with mauve veins and a yellow centre.
Description: This is a terrestrial species with linear to spathulate leaves. Flowers are 10–20 mm (0.4–0.8 in) across and appear in groups of two or more, on a 5 cm (2 in) scape. Traps are 1.5 mm (0.06 in) wide.

This is one of the easiest and prettiest of all *Utricularia* to grow—it multiplies freely and will flower all year in a pot sitting in a saucer of water on a sunny window sill.

Utricularia aurea x muellerii

Summit trail to Mount Kinabalu, Borneo.

Field Trips

In order to better grow, enjoy and find carnivorous plants, I have described some of the areas in which certain genera can be found. They include alpine New Zealand with its *Drosera* and *Utricularia*, temperate North America, with its Venus Fly Trap, *Drosera* and *Pinguicula*, Mount Kinabalu in tropical Borneo with *Nepenthes* and *Drosera*, Mount Roraima in Venezuela with *Heliamphora*, *Drosera* and *Utricularia*, and Western Australia with its *Cephalotus*.

Key

1. Albany, Western Australia
2. Anglesea, Victoria, Australia
3. Arthur's Pass, New Zealand
4. Mount Kinabalu, Borneo
5. Mount Roraima, Venezuela
6. North America

ALBANY, WESTERN AUSTRALIA: *DROSERA* AND *CEPHALOTUS*

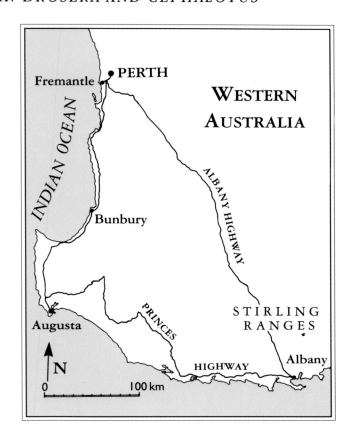

I set out from Perth in mid December at 3 pm to travel inland on Highway 95, the Albany highway. Albany is the oldest town in the state of Western Australia, dating back to 1826. On the highway to Albany, through the huge karri forests where trees reach up to 100 m (330 ft) tall, there are signs warning of kangaroos and the risk of bushfire. By 6.30 pm the Stirling Ranges had appeared on the left. With a highest peak 1042 m (3000 ft) above sea level, they contrast with land the other side of the ranges, towards the coast, which is incredibly flat. This fertile dairy cattle land is *Cephalotus* country. While *Cephalotus* appears throughout this whole region, it is most prolific 5-20 km (3-12 m) inland along the coast and 100 km (62 m) either side of Albany.

By 8 pm, having travelled some 400 km (248 m), I arrive at Albany. The sea breezes provide quite a cool night even in summer; as it becomes darker, the air becomes quite damp. Early the next morning I drive east toward King River, 10 km (6 m) from Albany. Much of the land around Albany has been cleared and drained for raising cattle, but where the natural vegetation exists, wild orchids and banksias are prevelant. As I turned off the highway onto a dirt road that led to a cattle property I expected to see *Cephalotus follicularis*, an endangered carnivorous plant species, protected under the Convention on International Trade in Endangered Species of Wild Fauna and Flora since 1979.

Stirling Ranges, Western Australia.

The area, complete with stream, is quite thick with sedges, myrtle (*Beaufortica sparsa*) and tea tree (*Leptospermum firmum*). The ground felt soggy as I walked towards the stream, and I noticed a few small scattered *Cephalotus* plants, some reddish burgundy, others green; some with pitchers broken by the cattle. There is a stark contrast between the short grazed areas and wild bushland. In the more natural and wild areas lush clumps of dark burgundy pitchers were everywhere: fifty or more plants stood in an area less than a square metre in size with smaller shoots emerging through the black peaty soil. The humus-rich soil also contains clumps of *Drosera hamiltonii* with large pink flowers, each clump having about ten to twenty plants, although some are scattered about singly and in pairs.

Amongst the burgundy pitchers of *Cephalotus* are a few 'normal' ovate leaves, also tinged with burgundy. These leaves are about 5 cm (2 in) long and 2 cm (1.75 in) wide, thick and waxy in appearance. They would have reached maximum size about a month previously and were now beginning to die off. New leaves appear in April, when they will dominate the plant. One plant I found had a 'primitive' pitcher that was half leaf and half pitcher, which looked like a curled up leaf with a half lid.

The peak of the insect season was due to arrive in another month, and the ultimate size of the pitchers corresponds to conditions in this period. All the larger pitchers sit on the ground with their lids open, waiting for insects. The lids are very hairy with dark burgundy ribs separated by tissue paper-like windows (areolae). Bending the lid back reveals an interior wall shiny and pinkish white near the rim, with small black dots indicating the small glands further down. At the very bottom is a blackish brown 'soup' of insects such as ants, mosquitoes, moths, flies and beetles. In the crown of some plants what appears to be a white, hairy new pitcher is developing. A few plants have large pitchers that have not yet reached the ground and these pitchers are still held shut, and contain a clear liquid. Green hairy flower stems emerge from the crowns: they are tall, up to 10 cm (4 in), and still not mature. Greener plants further into the scrub have even taller flower stems up to 25 cm (10 in) high. All these flowers will have opened by the end of the following month to reveal small white waxy flowers, which finish seeding a month later. Last year's brown withered scapes are still present on most plants. Continuing through the scrub towards the stream, the vegetation height drops to 60 cm (2 ft) and less; on the sloping banks the soil has more sand mixed with humus. As the vegetation thins the pitchers are reddish burgundy. Pitchers on the bank are so dark they appear almost black. Closer inspection reveals that there are ants trailing up into the pitchers. Not all the ants are trapped: some have found food and are heading back to their nest.

Many large white flowers of *Drosera myriantha* can be seen through the grasses, with their yellow green stem and leaves. This *Drosera*, which reaches a height of 25 cm (10 in), blends in with the grasses and uses them to cling to and climb.

Next, I took the coast road back to Perth via Augusta, the western most point of Australia, where the Indian Ocean and Southern Ocean meet at the north western limit of *Cephalotus* country. Along the coast are sheer cliffs of limestone and huge granite outcrops that alternate with white sandy beaches. River inlets cut into the coast and it is along the streams that feed into these inlets that *Cephalotus* can be found. Towards Augusta, *Cephalotus* seems to grow in a sandier, drier soil which, together with temperatures of 36°C (97°F) that day, resulted in the lids of the *Cephalotus* being closed to prevent evaporation of the contents of the pitcher.

The area from Albany to Augusta is also rich with *Drosera* and over twenty species can be found, including *D. bulbosa*, *D. stolonifera*, *D. platypoda*, *D. subhirtella*, *D. myriantha*, *D. menziesii*, *D. macrantha*, *D. gigantea*, *D. pygmea*, *D. pulchella*, *D. platystigma*, *D. paleacea* and *D. dichrosepala*. The area is also home to *Utricularia*, including *U. menziesii*, *U. hookeri*, *U. volubilis*, *U. simplex* and *U. multifida*. Apart from carnivorous plants hundreds of native wildflowers are still in flower, although most flowered a month before. In the swamps a further 250 km (155 m) north towards Pinjarra on the Murray River, I found *Byblis gigantea* in full flower. It was once thought this plant could catch rabbits: while

Cephalotus follicularis

this is hard to believe, seeing *B. gigantea* so densely packed you could easily imagine that a small bird would have little chance of extricating itself from this sticky mess if it happened to fly into it.

This short stretch of Western Australia contains many interesting carnivorous plant species. If you cannot make the trip down to Albany from Perth, however, you should at least drive through Kings Park, a 400 ha (988 acre) natural park in the middle of the city, which has many of these species in natural surroundings.

ANGLESEA, VICTORIA: SEARCHING FOR *DROSERA*

So diverse are *Drosera* in Australia that it is worth visiting a lowland area where a variety of species are to be found. Anglesea, Victoria, in the south east of Australia is such an area. It is a small seaside town, 100 km (62 m) west of Melbourne. The best place for carnivorous plants is 10 km (6 m) inland from the coast: here the vegetation is open forest, with Eucalyptus trees growing to a height of 15 m (50 ft). The annual rainfall is 300 mm (2.75 in) peaking at the end of winter, when the area can receive 80 mm (3 in) or more of rain, when the average temperature is 10°C (50°F).

The best time to visit the area is early spring, when the ground is still damp from winter rains and many flowers have bloomed—this is also before the tuberous *Drosera* revert to their tuberous state at the onset of summer. The best time of day to observe *Drosera* is the early morning, when droplets on the tentacles are bulging with liquid and the flowers are open. A few hours on and flowers close and the droplets disappear from plants exposed to full sun. This area has been protected for many years, and you must beware of kangaroos (they graze on scrub grasses, which grow up to 60 cm (24 in) tall). This grass provides filtered light for many *Drosera*. Anglesea is prone to bushfires, which scar the tall trees and burn back the scrub. If you visit this

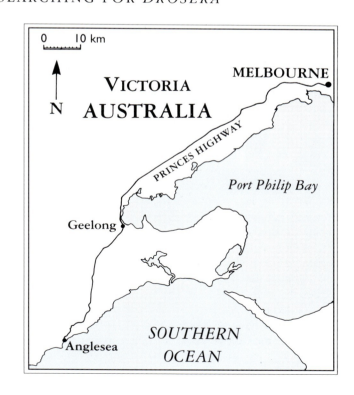

Anglesea, Australia.

area a year after a bush fire you will see bright green growth emerging from the burnt tree trunks and major regrowth from grasses and bulbous plants.

So prolific are the *Drosera* here that you will see them immediately. There is *D. peltata*, for example, which grows in fairly flat land in damp soil of clay and sand. It usually grows to a height of 10–30 cm (4–12 in), although you may see a few plants that are 60 cm (24 in) tall. The basal rosettes on these plants wither away as the flowers appear (once, plants with basal rosettes were thought to be different species). Flowers appear during spring; where the temperature is warmer closer to Melbourne, the flowering season may begin in August and finish in November, or may not begin until October closer to the coast.

The above-average height of *D. peltata* in Anglesea is possibly due to the fires in the area. It may also explain the vigour of these plants, which spring straight back as you brush past them (there are so many plants growing in the area that it is difficult to avoid stepping on them).

Just a few yards away is another tuberous *Drosera*, *D. whittakeri*, which emerges three or four weeks before *D.*

Drosera binata *at Anglesea.*

peltata and consequently also enters dormancy before it. *D. whittakeri* appears in clumps of twenty or more in open spaces between grass tussocks and is often covered by the leaves of eucalypts. From the way it appears in clumps, with smaller, younger plants radiating from older, larger plants you get the impression that seed survival in the wild is low and vegetative propagation by tuber is the superior method. It appears dark red in full sunlight in the open scrub (where vegetation is less than 20 cm (8 in) high). Where there are tall trees and the scrub has thickened and grown to a height of 60 cm (24 in), *D. whittakeri* is bright green. Whether in full sunlight or in shade, this species is always about 3–4 cm (1–1.5 in) in diameter.

Every time I see *D. whittakeri* I know that *D. peltata* is not more than a few metres away. Strangely, the reverse is not true. *D. whittakeri* is native to only two states of Australia, Victoria and South Australia, whereas *D. peltata* is so hardy it grows in every state in Australia. These are not the only *Drosera* growing in the area, and you are also likely to find *D. binata*, *D. glanduligera* and *D. pygmaea*.

ARTHUR'S PASS NATIONAL PARK, NEW ZEALAND

Arthur's Pass National Park is on New Zealand's South Island, on the east side of the Southern Alps. It links the east and west coast and is often 10° (50°F) or more colder than the north island, receiving twice the rainfall. It provides a perfect habitat for alpine and subalpine plants. It is quite an experience to see the change in vegetation from lowland, through montane to subalpine and finally alpine species. The whole trek culminated, for me, in the sight of *Drosera arcturi*, *D. spathulata* and *Utricularia monanthos* all clumped together in a natural bog. The easiest way to see these plants is to drive to the summit, step out of your car (the plants are not more than 100 m (330 ft) away from the car park), and drive back home. But to do this is to cheat yourself of the knowledge and beauty the landscape has to offer. A few hours experiencing and understanding the natural habitat of a plant you are trying to grow is worth more than any description and will, in the long run, save you months of heartache trying to duplicate the right environment for growing the plants.

The best time to travel to Arthur's Pass is summer, specifically January. At this time all the carnivorous—and many other—plants are in flower. Insects and wild fruits

Arthur's Pass, New Zealand.

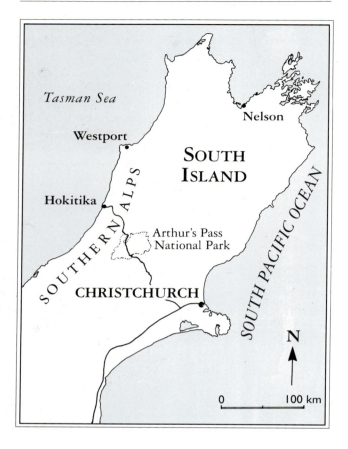

Montane vegetation is the level of vegetation that exists in altitudes 450–900 m (1480–2900 ft) above sea level. Below 450 m (1480 ft) the vegetation is considered lowland and is essentially grazing land for beef cattle and sheep. Where *D. spathulata* is found in the grazing lowland it usually has more than one flower, which appear three or more weeks before those on plants in the cooler mountain region.

As you continue to climb, the ground underfoot consists of leaf litter, broken branches and twigs, all rotted together. The rainfall in this area averages 5000 mm (196 in) each year, causing the soils to be leached and less fertile than the lowland. The rain and fog ensure that the soil is never dry, even during the extended summers. As you continue through this magical moss forest you may see wild deer or hear *kee-aah*, the sound of the Kea bird, an olive green mountain parrot that feeds on the wild berries.

Although Maoris may have first discovered Arthur's Pass in the seventeenth century, and used the wild berries and spawning salmon for food, it was named after the European Arthur Dudley Dobson who, in 1864, travelled through the alps looking for gold.

One of the attractions of this walk is Bridal Veil Falls. You know you are approaching when you hear the rumble of water then, around a bend, tonnes of clear water cascade in sheets straight down, to join Bridal Veil Creek. The spray of water increases humidity around this area and as a result there are many different types of ferns, some growing in

abound, and so, consequently, does the natural bird life. Check the weather before you set out as, even during summer, fog and rain on a freezing mountain with permanent snowfields on its summit can be very unpleasant.

Our trek starts at the main township near Arthur's Pass railway station, which in winter is covered in snow and looks like a Swiss chalet. The walk to the summit takes three or four hours travelling up the U-shaped Bealey Valley. The Bealey Valley was formed by a glacier. It's interesting to note that with such massive land movements *Drosera spathulata* is the only carnivorous plant found in the highlands as well as the lowland. All three carnivorous plants found on Arthur's Pass are also found in Tasmania, Australia, where they were first named. The three land masses—Tasmania, New Zealand and Australia—were joined millions of years ago.

On a warm sunny day the sky is clear, the air is fresh, the maximum temperature is 15°C (59°F) (but it will fall as low as 5°C (41°F) overnight), providing a pleasant walk up the mountain.

Five minutes into the journey you will cross Bealey Footbridge, then you are into the dense beech forest where almost everything is covered with spongy green sphagnum moss or pale green lichen. In some sections the canopy is so thick only limited light reaches the ground, but generally filtered light does break through this montane forest.

Arthur's Pass, New Zealand.

Arthur's Pass, New Zealand.

thin cracks in rocks. It is so moist that the sphagnum moss, when pressed, springs back like a mattress.

As you continue on, you cross the Bridal Veil Stream and get closer to the summit. The trees become shorter and scrubbier, the canopy drops (eventually to waist level) and frosts and snowfall are more frequent. The temperature can drop so low that waterfalls freeze. This area still experiences avalanches and so, despite the pleasant surroundings, this higher sub alpine (about 920 m (3020 ft) above sea level) zone can be harsh on trekkers and plants. Where the sun hits your body it is quite warm, in the shaded areas it is quite cold—the trick is to keep moving.

Eventually the forest path stops and you join the roadway. The area has only low scrub vegetation, little higher than 30 cm (12 in), which means most plants receive full sunlight and you can see for quite some distance.

Not far along the walk wooden planks are suspended above what appears to be very damp soil. This is the bog you have come to see. Planks prevent you sinking into peat moss, and also preserves the bog. The track climbs onto a rocky moraine ridge that overlooks the Otiro Valley—it is an amazing view looking down into a bog 200 m (660 ft) long and 100 m (330 ft) wide. Bog lands were formed as a result of glacial movement and they contain tarns. A tarn is a low lying water hole nestling in a bog. The star-shaped cushion plants (*Donatia nozae-zelandiae*), surround the tarn and, when they cover the whole of the bog the area is known as a 'cushion bog' or 'herb moar'. Other plants that exist around the tarn are sphagnum moss and red tussocks (*Chionochloa rubra*), named because of their reddish-brown base. Red tussocks are generally a good indicator of carnivorous plants, as they too exist in acidic, poorly-drained peat soil.

Continuing along the boardwalk as it descends, you will see flat swampy land and the first few *Drosera* nestling among cushion plants. Continue down and you will discover a tarn surrounded by *Drosera arcturi*, *D. spathulata*, and *Utricularia monanthos*—any uncertainty about finding carnivorous plants vanishes and you dart from one plant to another. Once you have recognised the general plant outline and the bright coloured flowers, you can step back and see that the whole area is covered with *Drosera*.

D. arcturi, from the latin *Arcturus*, Arthur, was identified in Mt. Arthur, Tasmania and occurs in areas south of latitude 40°. It is commonly called the Alpine Sundew. From a distance it looks like thin pinkish rustic fingers projecting from the green background. They cover an area of at least 50 m^2 (540 ft^2) with more than 5000 *Drosera* plants. Generally the leaves grow to 3 cm (1.2 in) in length, but they can reach 6 cm (2.4 in), when there is some shelter from the elements, and over twice that in a particularly good season. It presents a single, small white, five-petalled flower that rises just above the leaves in mid-summer (January). In a good season, two flowers may appear. Seed pods provide 25 or more small seeds and, to duplicate winter conditions, they should be refrigerated for three months prior to planting. Also, during winter a small winter nesting bud is formed and is covered by new short leaves and old withered ones, to provide protection during the harsh season. Although *D. arcturi* can flower from November to March, the best time to see it is in January, when *Utricularia monanthos* is also in flower.

U. monanthos, also found amongst the cushion plant, is readily identified by the bright purple flower, with a yellow spot in the centre (some varieties develop a white spot but this is rare). The flowers often rise above the *Drosera* on scapes 10 cm (4 in) long leading to the mistaken belief that they are the flowers of the closely-matted cushion plant, closed at this time of year. Commonly known as the Purple Bladderwort, *U. monanthos* flowers profusely after a dry spell. Its narrow linear to spathulate leaves are 5–25mm (0.2–1 in) long, and radiate out from the centre. They are deciduous in December and the tiny bladders are 1.5–2.5 mm (0.06–0.1 in) across, with fringed margins.

As if this were not enough, *D. spathulata* and the Pygmy Pine—the smallest conifer in the world—also grow amongst the cushion plant. *D. spathulata* is easily distinguished from *D. arcturi* by its short spoon-shaped leaves, and its overall size—the whole plant is only a quarter that of *D. arcturi*. It grows amongst the tussock and pokes through sphagnum moss and cushion plants, while elsewhere it stands out vividly on dark peaty soil, producing tiny red ringlets that look like red checkers. This is quite a versatile plant, that has spread from Malaysia to southern Japan and Australia, existing in lowland bogs as well as subalpine areas. At high altitudes it produces a single flower, often a few weeks later than its lowland counterpart. Unlike its lowland counterpart, highland plants usually die off during heavy winters, growing from germinated seeds the following season. *D. spathulata* produces white or pinkish petalled flowers from November to January. Each of the five petals measures 6 mm (0.25 in) in length. With the water table so close to the surface, the climatic conditions and the way the sphagnum acts like an enormous sponge, this area does not dry out during summer as it does in many other areas where *D. spathulata* grows. Originally found in Southport, Tasmania, it was named for the leaves shaped like a flattened spoon—hence the Latin *spathulata*.

To duplicate these conditions, to grow all three of these plants in a collection, the soil should be three parts peat moss and one of sand. Provide enough water that it can be topped up after total absorption every three or four days. If you have no trouble growing sphagnum moss year after year then consider growing all three species in it, otherwise stick to the peat mix. As these are alpine species keep plants on a low, cool bench if you expect temperatures to rise

On the road to Arthur's Pass, New Zealand.

above 30°C (86°F). During winter, if temperatures do not fall below 10°C (50°F), refrigerate *D. arcturi* and *U. Monanthos* as described for *Dionaea*.

Do not be deceived by the flatness and apparent stability of this tranquil countryside. The area in which the carnivorous plants grow is a large floating bog. If you leave the boardwalk you could sink, puncturing the fragile, growing, interwoven surface and you may not feel the bottom, so take care.

After you have spent backbreaking hours on your knees, observing these plants, continue up the path to Dobson's Monument: you have reached track's end, 920 m (3020 ft) above sea level. Further up, the flora ceases, with only the occasional lichen growing in the cold temperatures and brilliant sunlight. The mountains continue to climb, the highest being Mount Murchison at 2000 m (7800 ft), but over 25 per cent of the park exists below the 920 m (3020 ft) level.

There is always excitement tempered with concern when you first explore an area for carnivorous plants: you might not find the plants you have travelled hundreds of kilometres to see. Perhaps you have just walked past these rare plants or, worse, stepped on them, or the time of year is not right, or the information on which you have based all your hopes is not quite accurate. These concerns fade after experiencing one of nature's great attractions like Bridal Veil Falls.

Before you lose sight of the bog on your walk back, turn around. Try to imagine winter here—everything you can see covered in snow, frozen for months.

MOUNT KINABALU, BORNEO: FINDING *NEPENTHES*

It is impossible to travel through South East Asia looking for *Nepenthes* without visiting Borneo. Once in Borneo you must go to Sabah and climb Mount Kinabalu. Mount Kinabalu is undoubtably the botanist's Mecca, as this mountain, a tropical jungle, bridges the northern and southern hemispheres in terms of both plants and animals. Besides, the mountain contains more species of *Nepenthes* than any other area in the world, including some of the largest and most attractive pitchers. In fact, it holds species representing over half the families of flowering plants in the world, including *Rafflesia*, reputedly the largest flower in the world, measuring some 90 cm (36 in) in diameter.

Anyone interested in plants cannot help but be fascinated by the colour and variety on the mountain. When I climbed it for the first time I was physically tired but mentally exhilarated. From city centre to Mount Kinabalu takes a day by car, winding up the road to the base of the mountain. The trip is pleasant in this equatorial area, and you pass through many *kampongs* or villages, tropical plantations and roadside stalls of the Kadazan villagers.

Upon arriving at base camp we could see the top half of the mountain clearly for the first time. Soon it would

Mount Kinabalu, Borneo.

Rafflesia, one of the largest flowers in the world, on Mount Kinabalu.

be covered by afternoon cloud, obscured until early morning. This mountain that dominates Sabah and is at the back doors of its people was a magnificent sight. I was not sure whether I was up to such a climb. When I had told some of the residents in Sabah that I was aiming to climb the mountain, they said, 'the mountain! Good luck'. Many of the villagers had never climbed to the top, content to listen to tales of those who had made the climb.

It was an hour after arrival, and the mountain was covered in cloud. Soon the whole area was shrouded in a fine mist. Our small party, including a guide, set off at 8 am, with all our provisions in one pack that we carried in turn. I pondered how easy our trip was compared to travellers like Hugh Low who attempted it in 1851. He was the first scientist to climb the mountain and, with his party of 44 people he took 21 days to reach the summit. Very few of the Kadazan villagers had ventured onto the mountain before Low, for a variety of reasons, so the mountain was virtually untouched. Areas not easily accessible have still not been fully documented.

Some sections of the climb are like a set of stairs, with steps of tree roots worn smooth by the many climbers. The forest is alive with birds and animals, although the jungle is generally too dense to see them. In some places the forest is quite open, bathed in sunlight with dry and clay-like soil—in these areas small seedlings have a chance to grow to the canopy above. Other areas are dark and damp, with still warm air that allows mosses and fungi to thrive. A leaf falls to the ground and decays in a matter of weeks. I thought at the time that you could grow almost any plant in this warm humid climate all year round. Birdsnest ferns (*Asplenium nidus*) grow 6 m (20 ft) above the ground,

prevented from falling only by the fork of a tree. Common tree ferns (*Cyathea contaminans*) have fronds almost twice the size of a person and orchids are so plentiful that some are in flower all year round. As this area has no season except for the wet monsoon, many trees and shrubs flower all year round, relying on an apparently random and still mysterious time clock. I was amazed to see that this damp yet warm environment—so perfect for plants—was just what I have been trying to duplicate at home with heaters and sprinklers for all those years, all controlled by the latest technology. Since the jungle is one of the oldest of land environments it is no wonder that highly specialised plants such as *Nepenthes* have developed.

Nepenthes, commonly called Pitcher Plants, were used by many of the inhabitants of South East Asia to carry water. The pitchers are the shape of an urn and can hold water. Nature itself produces water in the pitcher, either from the plant or by atmospheric humidity condensing on the pitcher walls. *Nepenthes* is also called 'Monkey Cup' or 'Monkey Pot' as large apes, particularly orangutans, drink from them. Holding the pitcher like a china cup, these huge tropical apes drink with the grace of a participant at a tea party.

I have visited South East Asia and drunk from a developing *Nepenthes* pitcher that I opened. I can testify that it is quite a refreshing experience to drink the pale yellow liquid on a hot humid day—it tastes like mineral water, until one reaches the bottom where the taste is very strong. Even

Nepenthes burbidgae, *Mount Kinabalu.*

though the unopened pitcher contains no bacteria, I don't recommend drinking it as it can cause a gas build-up in the stomach. The sensation is quite pleasant with opened *Nepenthes*, as the underside of the lid contains droplets of water that are sweet, almost like honey.

After travelling for about two hours I sighted my first *Nepenthes*, *N. tentaculata*, hanging nearly 2 m (6 ft) up a tree and growing in a sphagnum clump. It is a small plant about 30 cm (12 in) long with six crimson pitchers no bigger than your thumb and with tentacles growing on the upper lid (hence the name). Looking around I realised we had walked into a mass of *Nepenthes*, both sides of the track populated with them. It was a magnificent sight. Some plants were growing on fallen trees covered with moss, others had taken root in the ground and stood 3 m (9.9 ft) into the air, and still others were beginning to fruit, with capsules open and seeds shed. Some pitchers on tall plants (over 2 m (6 ft)) had lost wings and tentacles and changed to a golden yellow as they grew closer to the sunlight. Others closer to the forest floor had light green pitchers, while others still have pitchers measuring as much as 15 cm (6 in) in length on leaves only slightly longer.

I was so overjoyed at seeing these *Nepenthes* in their natural surroundings that I felt if I did not see another pitcher plant, it would have been well worth the trip. We kept walking, finding more and more *Nepenthes*. Now we knew what to look for, spotting them was easier. Eventually though, the plants began to thin, and after altitudes of 2100 m (6800 ft) there were no more *Nepenthes* to be seen.

Interestingly, the mountain has a zonal climate. A tropical rainforest at the base changes to subtropical, temperate and eventually alpine regions at the summit. Many species exist in only one of these zones. As we left the upper tropical zone and entered the cooler moss forest zone, the oak trees, palms and ferns gave way to shorter trees, allowing more light to reach the forest floor. These dramatic changes in vegetation are surprising in an area with no summer or winter as such, only day and night temperature changes.

We continued to climb, resting frequently. The air seems thinner with a fine mist clouding the path ahead. I could feel the mystery surrounding this almost untouched land.

We walked for almost five hours when we discovered another little-known *Nepenthes*, *N. burbidgae*, native only to Mount Kinabalu. It is named after F.W. Burbidge who, in 1877, while working for the nursery of Veitch and Company, was only the sixth European to climb Mount Kinabalu. *N. burbidgae* has quite large pitchers that look surprisingly like porcelain jugs with dark red, bronze and brown patches streaking the beige background. The pitchers grow on a 1.5 cm (0.6 in) thick stem that crawls amongst trees and climbs towards the summit. Each leaf produces a tendril that grows until it meets a branch or vine to wrap itself around. Having found support it produces a pitcher and leaves it hanging, facing away from the plant with its

Nepenthes rajah, *East Mesilau River, Mount Kinabalu.*

lid open, waiting for prey. As the pitcher matures and fills with insects the plant produces another leaf, and the process continues. Large apes use these vines or lianas to climb trees, and fictitious characters use them to swing from tree to tree. Unfortunately, we only found two specimens of *N. burbidgae* in the area.

The next *Nepenthes* we saw was a very impressive *N. villosa*, at altitudes of 2500 m (8200 ft). It was a reddish, golden-tan colour. As I put a finger inside the pitcher to test the liquid depth and felt the sharp teeth on the rim I had to remind myself that this was only a passive carnivore. Further on I saw another plant, then another and another— the area was covered with *N. villosa*, some with pitchers so large you could fit your fist inside. The plant itself was short (no taller than 60 cm (24 in)) and stout, with 38 cm (15 in) long thick leaves like those of a rubber tree. Some of the pitchers nestled in the sphagnum moss close to the ground so that we could only see the lid and an opening amongst the tall thick grass (*Deschampsia flexudsa*), indicating a very acid soil. There were 30 or 40 plants either side of the track. Some had their yellow lids closed, with pitchers one third filled with water, while others were chocolate brown and black as they began to decay.

We continued up the track, wanting to climb higher and wanting to see more plants, knowing that these species would soon disappear. The vegetation was changing again, mosses gave way to lichens and the mist became thicker.

As we reach altitudes of 3000 m (9800 ft) the trees, sometimes growing between rocks, became more scrub-like, stunted and twisted like bonsai. The vegetation was sparse with no tree ferns or palms. The thin air made resting periods more frequent. As I climbed higher I found myself stopping every 15 minutes, wishing it was every five minutes. It was also quite cold, only 13°C (55°F). The ground was a lot drier by this stage and leaf mould and mosses were giving way to chunks of granite rock. I looked back and considered the distance I had travelled, and the trees below were only just visible through the fog. The trees had stopped, the ground was almost solid granite and to my delight I could see the summit camp ahead. This was where I would spend the night before setting off back down the mountain.

MOUNT RORAIMA, VENEZUELA

One of the most well-known of the 100 or more tepuis—high plateaus—in South America is Roraima, a large triangular outcrop that marks the border of Venezuela, Brazil and Guyana and is part of the Panaraima mountains. The nearest capital city is Georgetown, Guyana, from which point journeys traditionally begin. Now, however, journeys can begin in any of the surrounding cities, and you can fly to the summits of the tepuis by helicopter. If you choose to begin on the ground, there is an immediate sense of the tropics in Georgetown when you arrive, with its warm humid weather and a skyline dotted with palm trees. An interesting feature of the climate in this region is that there are usually two seasons repeated every six months. The yearly seasons are two wet and two dry. During the wet season the downpour (an average of 250 cm (100 in) a year) is short-lived and followed by immediate sunshine. So sunlit is this land that only about 20 days of the year experience less than full sunlight, with total yearly sunlight of 2100 hours. Clothing should be light as the temperature seldom falls below 24°C (75°F) and rarely rises above 33°C (90°F). This warm temperature, together with the rainfall and cool sea

Auyun tepui, Venezuela.

breezes, results in humidity levels of 78 per cent all year round. Even the nights are very pleasant with no great drop in temperature. Further inland the temperature changes become even less variable, perhaps not varying more than 1.6°C (3°F) from the coldest to the hottest months. Guyana is not all paradise, however, and the malaria mosquito can cause even the most determined carnivorous plant enthusiast to turn back.

The surrounding landscape is incredibly flat, and vegetation is lush with broad green leaves from trees such as the cannon ball tree (*Couroupita guianensis*), named because of the large light brown cannon ball shaped fruit that hangs down from the trunk on long thin stems. Trees like *Couroupita* respond to the seasonal changes by losing their leaves up to four times each year. Leaves are shed over just a few days, then within just a few more days, leaves grow to replace the fallen.

Once at Georgetown, and after taking malaria injections, the trip to Roraima takes about one day, crossing a number of rivers through the lush jungle of rubber trees to the flat savannah grasslands out of which the tepuis rise. Guyana, the land of waters, has many rivers, all lined with dense vegetation. These rivers eventually form the great Amazon River, containing both electric eel and the carnivorous piranha fish. Much of the water that fills these rivers originates in the mountain plateaus. The pure water cascades down the cliffs, producing magnificent falls such as Kateteur, which is higher than Niagara. While the extra water of the wet season makes this attractive, the best time to climb the plateaus is during the dry seasons of August to September and March to April, as after heavy rains the savannah becomes swampy and difficult to traverse.

Through the jungle, entering the savannah grasslands, you gain the first glimpse of Mount Roraima and the regal Mount Kukenan away to the left. As you travel through the grassy savannah towards the base of the mountain the vegetation changes, and the forest rises like a fence around the savannah. We approach the base of Mount Roraima through damp woods and tropical vegetation, and the temperature is a constant cool 15°C (59°F). The mountain looks like a flat table, often covered in fog. The dark red colour of the sandstone cliff 'table' is broken by white streamers hanging over it. These streamers are waterfalls of mist and rain that 'decorate' the land.

Genlisea roraime, *Mount Roraima*.

Brocchinia steyermarkii, *Mount Roraima*.

You will need to set out at 4 am to scale the mountain, so that by 3 pm you will be resting at the top. During the climb you find many areas are extremely wet from the spray from above, as if it is constantly raining. Natives call this mountain the 'mountain of rivers and the mother of storms'. As you pass the clouds, the temperature drops to −12°C (10°F), where it stays constantly, day and night.

At the summit the land is not as flat as it appears from below. Huge rock formations are scattered about, some up to 45 m (148 ft) high. Smaller formations look like frogs, mushrooms and wolves, and one looks like a monkey eating a giant ice cream cone. The mountains in the distance look like stepping stones in a sea of white fog.

On this bare, wind-blown rock plateau you wonder how any plants survive. The trees are stunted, the bushes are low and, except for the wind, there is total silence. But on this 390 square km (150 square miles) triangle there are sheltered areas that provide small oases in which very rare animals and plants are isolated from the rest of the world. In these oases mosses, ferns, rare orchids, bromeliads and palms flourish, along with *Heliamphora nutans*, *Drosera roraimae* and *Brocchinia reducta*. Another rare species is *Arundinaria schomburgkii* from which the indigenous peoples make their pipe dart poison called curare.

A visit in the wet season will show you many plants in flower. So damp is it that many fallen branches are covered in moss and instantly crumble when stepped on. The dampness comes from the fog that has been known to cover the summit for weeks. The fog and clouds in lower parts of the tepuis disappear in late afternoon with the warm sunlight, but this sunlight isn't enough to warm the ground and icy water in the pools at the summit—this is not a good place to spend the night, and all but the most hardy enthusiasts should head back in late afternoon.

NORTH AMERICA: SEARCHING FOR *DIONAEA*

If you, like eighteenth-century botanist John Bartram, were to search for *Dionaea*, you would need to travel to the flat coastal plains on the east coast of North America. The thin strips of sand, gravel, alluvium and unconsolidated deposits in grassy savannah, broken up by patches of swamp and long-leaf pine trees that drop their needles every other year (increasing the acidity of the soil) is ideal for this species. *Dionaea* is found in only two states, North and South Carolina, with their mild winters (averaging 10°C (50°F)) and warm, humid summers (averaging 27°C (80°F)). The high sunlight hours—2800 hours each year—and heavy rainfall (1.5 m (5 ft)) in the area makes for a humid climate with effectively no dry season. These conditions are ideal for *Dionaea*.

Once you have decided on the area to explore, you must look for mixed coniferous and deciduous forest vegetation that is swampy—take your gumboots! Areas that are known for rattlesnakes and alligators are also likely sites. It is in this area where the long-leaf pine forest ends and the wet evergreen bogs begin, an area known as Pocosin, that *Dionaea* will be found.

Look into the glistening sand at your feet and you will see (during May and June) small white flowers rising 30 cm (12 in) from the ground on slender green stalks. If you are looking in June and July all the petals will have turned black as the seed pods ripen. Out in the open in full sunlight, below the white flowers and in amongst the patchy grasses, you will discover bright red, tall, erect traps. They are in clumps of five to ten plants, then nearby there is a barren wet patch, then another clump. You will be on your knees, moving aside blades of grass, then you will straighten up, stand back, and find to your surprise many hundreds of clumps as you identify that distinct outline of the Venus Flytrap. Bend to ground level again and look inside the closed traps of *Dionaea*: you will find they have caught mainly ants and spiders, some grasshoppers and beetles

and just a few mosquitoes and flies. After five to ten days these feeding traps will begin to open and the wind will blow away any wings or skeletal remains.

You may be lucky enough to find *Dionaea* growing alongside *Drosera* (say, *Drosera capillaris* and *D. intermedia*) in wet sandy ditches beside the roadside. Nearby you may even find *Pinguicula* species such as *P. lutea* and *P. caerulea*. If the area is wet all year round you may also find other carnivorous plants such as *Utricularia purpurea* and *U. inflata*—truly a feast for the carnivorous plant enthusiast.

If you return to the area six months later *Dionaea* will be harder to find, as the coastal plain has a high water table and much flooding. Significantly, *Dionaea* often thrives in quite varied water levels. When plants grow in full sunlight the water level is quite high; when plants are shaded or almost covered with pine needles the water level is lower. Evergreen shrubs, small trees and sphagnum moss almost cover up *Dionaea*'s winter growth, producing leaves that are ground-hugging with fewer traps, which are also smaller and on shorter, wider leaves.

Swampy areas make observation of these plants uncomfortable for the explorer, but they ensure *Dionaea*'s survival. When areas have been drained the pine trees soon take over and the *Dionaea* are squeezed out, and if an area has not had a fire for ten years or more the shrubs and forest become very dense, effectively inhibiting the growth of *Dionaea*. Fires every one to three years help sharpen the edge between the Pocosins and the forest by eliminating the smaller trees. When fire is followed by rain it produces a soggy acidic soil, ideal for regrowth of *Dionaea* from its underground rhizome. *Dionaea* often provides the first flashes of green in a burnt black landscape.

Further inland climatic conditions are less favourable, soil changes to sedimentary rock and the pH levels increase, limiting the spread of *Dionaea* and containing it within this small area of the world.

Dionaea muscipula.

Glossary

This is a list of *some* of the terms used when discussing carnivorous plants. Throughout the text, terms have been explained as they first occur: this list, however, will serve as a useful point of reference.

Apex: top or end point.
Attenuated: diminishing in width.

Bladder: the sac-like trapping structure of *Utricularia*.
Bog: an area of permanently wet land, with a moist atmosphere, usually dominated by mosses and herbaceous perennials (also known as tarns).
Bract: modified leaf found along the flowering stem or below the calyx.
Bud: a mass of tissues that develop into a stem, flower, or branch.
Bulb: swollen underground section of the plant.

Calyx: collective term for the sepals.
Chasmogamous: flowers that are pollinated when open.
Cleistogamous: self-pollinating flowers.
Column: structure supporting the lid or hood of *Sarracenia*.
Corolla: collective term for the petals of a flower.
Cross-pollination: pollination from one plant to a different plant.

Endemic: specific to a particular region, and found nowhere else in nature.
Enzyme: substance that helps digestion.
Epiphyte: plant that lives on another plant, using it for support, but which is *not* parasitic.

Fenestration: translucent section, as appears on the hood of pitchers of *Darlingtonia* and *Sarracenia*.
Filiform: thread-like.

Gemmae: reproductive structures; bud-like structure that can produce identical plants to the parent plant.
Glabrous: without hair.

Heterophyllous: plants with different types of leaves (shape, size, function) at different stages.
Hibernaculum: winter resting bud.
Hood: lid-like appendage that hangs over the opening of a pitcher.

Inflorescence: floral section of the plant.

Lanceolate: narowing to a point.

Midrib: central vein of a leaf.
Mucilage: a glutinous substance secreted by some plants.

Panicle: a compound raceme.
Pedicel: small stem supporting a single flower.
Peltate: umbrella-shaped.
Peristome: teeth-like structure on the rim of pitchers, as with *Nepenthes*.
Peristomes: teeth-like structures.
Perlite *see* **Scoria**.
Petiole: leaf stalk.
Phyllodia: winter leaves; broad petioles without leaf blades, which function as leaves.
Pitchers: tubular leaves, such as those on *Cephalotus, Darlingtonia, Heliamphora, Nepenthes* and *Sarracenia*.
Pubescent: hairy.

Raceme: an unbranched inflorescence.

Rhizome: horizontal underground stem, from which roots and branches grow.

Scape: stem or stalk with no leaves, that may bear flowers or bracts.
Scoria: growing medium; silicate particles.
Sessile: being attached directly to the branch of the plant, with no stem.
Spathulate: spatulate; spoon-shaped.
Sphagnum moss: a variety of moss that grows in temperate bogs and is a useful growing medium; also known as peat moss and bog moss.
Stolon: a runner; a stem that grows on the surface of the soil and often produces roots.
Stratification: the placement of seeds between layers of moist sand and peat, after which they are exposed to low temperatures. This process replicates dormancy, and is required by some species before germination occurs.

Tendril: a modified leaf or branch that curls around nearby plants or structures, to help support the plant.
Trigger hairs: spike-like structures that, when stimulated, trigger a response, eg inducing trapping in species such as *Aldrovanda, Dionaea* and *Utricularia*.
Tuber: a fleshy underground stem that stores food and water and has a propagatory function.
Turion: hibernaculum produced by many water plants, including some *Utricularia*.

Vermiculite: mineral residue from mica and biotite; used as a growing medium.

MONTHLY CALENDAR

	SUMMER		
NORTHERN HEMISPHERE SOUTHERN HEMISPHERE	JUNE DECEMBER	JULY JANUARY	AUGUST FEBRUARY
ALDROVANDA	watch for algae	keep constant water level all year flowers appear	
BROCCHINIA reducta	new pups and flowers; pot up; keep pups damp	pot up pups; keep damp	keep damp
BYBLIS gigantea	dormant; keep dry	dormant; keep dry	dormant; keep dry
BYBLIS liniflora	flowering begins; keep soil damp	cross-pollinate; keep soil damp	maximum size; keep soil damp
CATOPSIS berteroniana	increase watering	increase watering	keep well watered
CEPHALOTUS	flowers continue to grow	cross-fertilise; keep damp	growth peaks; keep damp
DARLINGTONIA	rhizome grows; keep damp	rhizome growth continues; keep roots cool	take cuttings; keep roots cool
DIONAEA	rapid growth continues as flowers die; sit pot in water	traps turn red; sit pot in water; take cuttings; plant seeds	growth continues; sit pot in water; thin out seedlings
DROSERA non-tuberous	water by tray	water by tray	growth continues
tuberous	dry soil; dormant tuber; repot if necessary	dry soil; dormant tuber; repot if necessary	dry soil; tuber dormant
DROSOPHYLLUM	decrease water for 3 weeks	never repot	growth continues
GENLISEA	stand pot in water and keep warm all year		
HELIAMPHORA	moist sphagnum; mist regularly	avoid intense sunlight and heat; mist regularly	avoid intense sunlight and heat; mist regularly; some ssp flowering
NEPENTHES	collect seed; keep plant warm	collect seed; keep plant warm	shade some species; keep plant warm
PINGUICULA Southern species	provide shade; keep moist; sow seeds	provide shade; keep soil moist; some species dormant	growth peaks; keep soil moist
Northern species	flowers appear	flowering continues	growth slows
SARRACENIA	sepals wither; seeds developing; sit plant in water	flower pod shrinks; sit plant in water	collect seed; sit plant in water
TRIPHYOPHYLLUM	increase watering; seeds germinate; keep warm and humid all year; glandular leaves continue to grow		
UTRICULARIA terrestrial–nontuberous	collect seed, flowering continues	keep moist all year take cuttings	stand plant in water
terrestrial–tuberous	keep tuber dry	keep tuber dry; separate and	keep dry
aquatic	watch for algae	keep water level topped up repot if necessary	

AUTUMN/FALL

SEPTEMBER / MARCH	OCTOBER / APRIL	NOVEMBER / MAY	NORTHERN HEMISPHERE / SOUTHERN HEMISPHERE
keep constant water level all year			**ALDROVANDA**
keep damp	keep damp	keep damp	**BROCCHINIA** reducta
begin to water	keep soil damp	shoots appear	**BYBLIS** gigantea
flowers wither; keep soil damp	seeds begin to ripen	collect seeds	**BYBLIS** liniflora
keep well watered; flowers emerge	keep well watered; flowers emerge	keep well watered; flowers continue to grow	**CATOPSIS** berteroniana
rhizome thickens; new shoots appear	pitchers redden	pitchers achieve full size	**CEPHALOTUS**
collect seeds	take leaf cuttings	decrease watering	**DARLINGTONIA**
growth continues; sit pot in water	maximum size achieved	growth slows down	**DIONAEA**
medium height	plant begins to shoot	seeds ripen after about two weeks	**DROSERA** non-tuberous
plant emerges from dormancy	shoots break ground	begin to water	tuberous
sow seeds; keep warm (above 10°C (50°F)	seedlings appear; keep warm (10°C (50°F)	decrease watering; keep damp	**DROSOPHYLLUM**
stand pot in water and keep warm all year			**GENLISEA**
ensure ample light; mist regularly	ample light; mist regularly	growth slows	**HELIAMPHORA**
well-drained soil; keep plants warm	keep plants warm and soil well-drained; provide ample light	well-drained soil; keep plants warm; ample light	**NEPENTHES**
growth slows; allow drainage	dormant; decrease water; damp soil	dormant; decrease watering, keep damp	**PINGUICULA** Southern species
growth slows	winter buds emerge; old leaves decay	dormant; decrease watering, keep damp	Northern species
maximum size, decrease water	growth slows; sow seeds; S. oreophila develops phyllodia	pitchers wither	**SARRACENIA**
keep warm and humid all year			**TRIPHYOPHYLLUM**
keep warm and humid all year			**UTRICULARIA** terrestrial–nontuberous
keep dry	keep dry	new shoots appear; begin to water	terrestrial–tuberous
	keep water level topped up		aquatic

	WINTER		
NORTHERN HEMISPHERE SOUTHERN HEMISPHERE	DECEMBER JUNE	JANUARY JULY	FEBRUARY AUGUST
ALDROVANDA	keep constant water level all year		
BROCCHINIA reducta	decrease water level in conditions below 10°C (50°F)	decrease water level in conditions below 10°C (50°F)	decrease water; beware of temps below 10°C (50°F)
BYBLIS gigantea	growth continues	growth continues; take cuttings	flower buds begin to grow
BYBLIS liniflora	plant begins to wither; decrease water	plant dies off	plant dies off
CATOPSIS berteroniana	flowers self pollinate; decrease water	seeds form; decrease watering	new shoots and roots appear; main plant begins to die
CEPHALOTUS	non-carnivorous leaves appear; growth slows	growth slows; decrease water	keep damp
DARLINGTONIA	allow pot to drain and dormancy to begin	divide; refrigerate rhizome if necessary	drain well; repot if necessary
DIONAEA	dormancy begins, refrigerate rhizomes if necessary	dormancy; keep plant cool and wet; some leaves die back	dormancy; keep cool and damp
DROSERA non-tuberous	gather seeds, plant approaches dormancy; collect gemmae	some leaves decay, others stop growing, depending on species	damp soil only; collect gemmae
tuberous	plant begins to fill out and flower in some species	plant and flower growth continues; keep damp	maximum size for some species new tubers develop; keep damp
DROSOPHYLLUM	keep warm above 10°C (50°F)	growth begins; keep warm, above 10°C (50°F)	growth continues; keep warm above 10°C (50°F)
GENLISEA	stand pot in water and keep warm all year		
HELIAMPHORA	keep warm and damp	keep soil warm and damp	keep warm and damp; divide if necessary
NEPENTHES	ensure well-drained soil, warmth, full light; growth slows	keep well-drained; maintain winter temp above 15°C (59°F)	keep soil well-drained; keep warm in full sun; flowers appear
PINGUICULA Southern species	dormant; some species form resting buds	dormant	gradually increase water; repot if necessary
Northern species	sow seeds; roots decay in some species	repot if necessary; problem period; remove gemmae	problem period
SARRACENIA	dormancy; allow to drain	dormancy; trim off dead pitchers	dormant; divide plant and repot; flower buds appear
TRIPHYOPHYLLUM	keep warm and humid all year		flowers appear on older plants
UTRICULARIA terrestrial–nontuberous	new shoots appear		growth continues
terrestrial–tuberous	shoots continue to grow; drainage required	stand pots in water	flowers begin to appear
aquatic	keep water level topped up all year		

This diary should only be used as a guide. Some species differ greatly from the genus characteristics outlined. If you live in a warm climate, plant cycles will be one month or more ealier than mentioned above, dormancy may be shorter and winter leaves may not emerge for three or more months. Plants propagated by tissue culture should generally

SPRING

MARCH / SEPTEMBER	APRIL / OCTOBER	MAY / NOVEMBER	NORTHERN HEMISPHERE / SOUTHERN HEMISPHERE
repot, divide	watch for algae	watch for algae	**ALDROVANDA**
fertilize; pot up pups	fertilize; pot up pups	fertilize; pot up pups	**BROCCHINIA** reducta
flower buds continue to show	flowers open	cross-pollinate	**BYBLIS** gigantea
plant seeds	seedlings sprout	growth continues; keep wet	**BYBLIS** liniflora
remove pups; pot up	remove pups; pot up; main plant nearly dead	increase watering	**CATOPSIS** berteroniana
keep damp; repot; take cuttings	increase watering	flower stem begins	**CEPHALOTUS**
flowers emerge	flowers grow	some plants still in flower	**DARLINGTONIA**
remove rhizome from refrigerator; repot; divide	sit plant in water; growth begins; take leaf cuttings	flowers emerge; cross-pollinate; few small traps emerged	**DIONAEA**
sow seeds; repot; increase water; flowering begins	growth begins; seeds sprout; sit plant in water; flowering cont.	flowering continues; growth continues; water by tray	**DROSERA** non-tuberous
decrease water some spp; secondary tubers develop	tubers continue to grow; leaves absent	no growth above soil	tuberous
growth slows	growth slows	flowers emerge; growth slows	**DROSOPHYLLUM**
stand pot in water and keep warm all year		flowers emerge	**GENLISEA**
flowers emerge most species; mist regularly	growth increases; mist regularly	avoid intense sunlight and heat, mist regularly	**HELIAMPHORA**
sow seeds; keep warm; flowering continues	seedlings begin to appear; two plants necessary for fertilisation	flowering continues; shade for some species; repot if necessary	**NEPENTHES**
growth begins; sit in water; take leaf cuttings	growth continue; keep moist; cuttings shoot; flowers appear	flowering continues	**PINGUICULA** Southern species
seeds germinate; new growth begins	growth continues; shoot gemmae	watch for algae	Northern species
flower buds appear; increase water; seedlings appear	flowers continue to appear; sit plants in water	pollinate now; sit plant in water	**SARRACENIA**
flowers appear on older plants	sow seeds; flowers appear on older plants	increase water; sow seeds; glandular leaves produced	**TRIPHYOPHYLLUM**
new leaves appear; sow seeds	increase watering; repot if necessary	flowering of many species	**UTRICULARIA** terrestrial–nontuberous
collect seeds; growth slows	decrease watering	allow to dry out	terrestrial–tuberous
flowers emerge; keep water level topped up	flowering continues; watch for algae	watch for algae	aquatic

be planted out about one month after the time specified for planting seeds, except where plantlets are fairly mature, when they should be planted out approximately two months after planting time. Consult genus and species descriptions, as well as sections on cultivation and propagation, for more specific information.

Cultivating Guide

Plant	Origin	Growing Medium	Watering Method	Light	Propagation Method	Humidity %	Tempeature range °C/°F	Difficulty	Insecticide	Fungicide
Aldrovanda	Widespread	Peat Water Ph. 4-6	Tu	M	S,Sc		20-30 (68-86)	Care required		
Byblis										
gigantea	Australia	1P₂+1P 1P+Sa 1Sc+1P	Wd,D	M-F	S,Tc,Sc	50-70	10-40 (50-104)	Difficult	Wo	To,B,F,Tr
liniflora	Australia	2P+1Sa	T	M	R, S, L	50-70	15-40 (59-104)	Easy	Wo	To,B,F,Tr
Cephalotus	Australia	3P+1V+2Sa 3P+1Sa	T(Wi) Wo(Su)	F-S	L,R,S,Tc	60-80	10-35 (50-95)	Care required	Wo	B,F,Tr
Darlingtonia	North America	3P+1V 1P+1S S	Wd,D	F	L,R,S,T	50-70	5-30 (41-86)	Care required	Wo,P,Ma	D,BF,Tr
Dionaea	North America	3P+1V P 3P+1S	T	M-F	L,R,S,Tc	50-90	5-40 (41-104)	Easy	A,P	B
Drosera										
Non-tuberous	Widespread	tropical 3P+2Sa; S subtropical 3P+2Sa	T	F-S	L,S,Tc,La,G	50-90	5-40 (41-104)	Easy	A,We,Me	F,Tr
Tuberous	S. Australia	3P+2Sa	D,Wd	M-F	S,R	50-70	0-30 (32-86)	Care required	Ma,Wo,Me	B,F,Tr
Drosophyllum	SW Spain, Portugal Morocco	2P+1Sa 3Sa+1P+2m	D,Wd	M-F	S,Tc	50-70	Sa(10-30) (50-86) Wa(10-12) (50-54)	Care required	Wo, Me	B
Genlisea	South America and Africa	P	T	M-F	S,Sc	60-80	10-30 (50-86)	SA: Difficult Africa: Easy		
Heliamphora	South America	S,P+Sa+Lm 4P+1Pe	Wd,D	F-S	S,R	60-100	10-25 (50-77)	Care required	Ma,P	B,F,Tr
Nepenthes										
Lowland	S.E. Asia, etc.	S,P+Sc, Cb	Wd,M	M-S	S,Tc,Sc,La	70-100	15-35 (59-95)	Care required	Wo,P	B,F,Tr
Highland	S.E.Asia, incl. Sri Lanka, India, Madagascar	S,P+Sc, Cb+P	Wd,M	M-S	S,Tc,Sc,La	80-90	8-27 (47-81)	Care required	Wo,P	B,F,Tr
Pinguicula										
Homophyllous	Sthn USA, Europe, Cuba, Haiti, Cyprus	P+Sa	T	M-F	S(L)	60-80	5-30 (41-86)	Care required	A	T
Heterophyllous	Mexico	Ptle 2P+1Sa+1Pa	Su (T) Wi(WD)	M-F	S,Lc	50-80	10-30 (50-86)	Easy	A	T
Hibernacula	Europe, Asia, North America	P+Sa+Le	T	M-F	S,G	60-80	S (10-30) (50-86) Wi (1-4) (33-40)	Care required	A	T

Plant	Origin	Growing Medium	Watering Method	Light	Propagation Method	Humidity %	Tempeature range °C/°F	Difficulty	Insecticide	Fungicide
Sarracenia	North America	P+Sa,P	T	M-F	S,R,Tc	50-90	-5-35 (23-95)	Easy	Ma,A,Wo,P	B,F,Tr
Triphyophyllum	West Africa	2Sa+lP	Wd,D	F-S	S	50-90	15-30 (59-86)	Cultivated, but difficult	Ma	F
Utricularia										
Aquatic	Widespread	Peat Water PH 6	Tu	F-S	S,,Sc	60-80	15-30 (59-86)	Care required		
Terrestrial non-tuberous	Widspread	P+Sa	T	M-F	S,Sc,,L	50-70	5-35 (41-95)	Easy	Wo,Ma,P	B,F,Tr
Terrestrial tuberous	Widspread	Lm+Cb+Pe P+S	T	M-F	S,R	50-70	5-35 (41-95)	Care required	Wo,Ma,P	B

KEY

Growing medium

- P   Peat moss
- S   Sphagnum Moss
- Sa Sand
- V   Verniculite
- Lm Leaf mould
- Sc Scoria
- Cb Chipped bark
- Pe Perlite
- le Limestone

Watering methods

- Mi Misting
- Su Summer   where species require dramatic
- Wi Winter   variations in watering.
- T   By tray (containing 2–3 cm (0.75–1.25 in) water)
- Tu (for aquatic species) keep water levels topped up and at a constant level.
- Wd Well-drained soil
- D   Damp

Light

- M   Maximum sunlight
- F   Filtered light
- S   Full shade

Propagation methods

- G   gemmae
- L   Leaf cuttings
- La Layering
- R   Root cutting
- S   Seed
- Sc Stem cutting
- Tc Tissue culture

Fungicides

- B   Benlate or Benomyl
- To Topsin
- D   Dithane
- F   Fongarid
- Tr Truban

Insecticides

- Wo White oil
- Ma Malathion = Maldison
- K   Kelthane
- A   Acetelic
- Me Metasystox
- P   Pyrethrin

World Carnivorous Plant List

Following general scientific procedure all species are in *italics*. For ease of reference, straight species are shown in ***bold italics*** and subspecies, varieties and forms are in *italics* alone. For more detailed information and explanations of the structure of this table, abbreviations and conventions, please consult key at end of this list.

Aldrovanda
Family: DROSERACEAE

Aldrovanda vesiculosa L.—Europe, India, Japan, Africa, Australia

Brocchinia
Family: BROMELIACEAE

B. hectioides Schultes—Venezuela
B. reducta Schultes—Venezuela

Byblis
Family: BYBLIDACEAE

Byblis gigantea Lindl.—Australia
Byblis linifora Salisb.—Australia, New Guinea

Catopsis
Family: BROMELIACEAE
Sub-family: TILLANSIOIDEAE

Catopsis berteroniana Schultes—S. America, USA, West Indies

Cephalotus
Family: CEPHALOTACEAE

Cephalotus follicularis Labill.—Australia

Darlingtonia
Family: SARRACENIACEAE

Darlingtonia californica Torr.—California, Oregon

Dionaea
Family: DROSERACEAE

Dionaea muscipula Ellis ex. L.—E. USA

Drosera
Family: DROSERACEAE

D. acaulis L. F.—S. Africa
D. adelae F. Muell.—Australia
D. affinis Welw. ex. Oliver—Tropical Africa
D. alba Phill.—S. Africa
D. aliciae Hamet—S. Africa
D. andersoniana Fitzg. ex Ewart. et White—Australia
D. androsacea Diels = *D. parvula*
D. x anfir Nagamato = *D. anglica x filiformis*
D. anglica Huds.—Europe, N. America, Japan
D. angustifolia F. Muell. = *D. indica*
D. annua Reed = *D. brevifolia*
D. x anpill Nagamato = *D. anglica x capillaris*
D. x anterm Nagamato = *D. anglica x intermedia*
D. arcturi Hook—Australia, New Zealand

D. arenicola Steyermark—Venezuela
D. ascendens Planchon.—Brazil = *D. villosa*
D. atra Col. = *D. arcturi*
D. auriculata Backh. ex Planchon.—Australia, New Zealand = *D. peltata*
D. banksii R. BR. ex DC.—Australia
D. barbigera Planchon.—S.W. Australia
D. bequaertii Taton—C. Africa
D. binata Labill.—Australia, New Zealand
D. brevifolia Pursh—N. America
D. bulbigena Morrison—Australia
D. bulbosa Hook.—Australia
D. burkeana Planchon—South Africa
D. burmanni Vahl.—Asia, Tropical Australia
D. burmanni var. *dietrichiana* Reichb.
D. caledonica Viell. ex Diels—New Caledonia
D. calycina Planchon.—W. Australia = *D. microphylla*
D. capensis L.—S. Africa
D. capillaris Poir.—N. and C. America, Northern South America
D. cayennensis Sagot ex Diels—Guyana, Brazil
D. cendeensis Tamayo et Croizat—Venezuela
D. chiapensis Matuda—Mexico
D. chrysolepis Taub.—Brazil
D. circinervia Col. = *D. auriculata*
D. cistiflora L.—S. Africa
D. collinsiae Brown et Burtt Davy—S. Africa = *D. burkeana* x *madagascariensis*
D. colombiana Fernandez-Perez—Colombia
D. communis St Hil.—Brazil, Colombia
D. compacta Exell. et Laundon—Angola
D. congolana Taton = *D. madagascariensis*
D. corsica Maire—Corsica = *D. rotundifolia*
D. cuneifolia L. ex Jackson—S. Africa
D. cunninghamii Walp. = *D. binata*
D. curvipes Planchon. = *D. madagascariensis*
D. curviscapa Salter = *D. aliciae*
D. dichrosepala Turcz.—Australia
D. dielsiana Exell. et Laundon—S. Africa
D. dietrichiana Reichb. = *D. burmanni* var. *dietrichiana*
D. dilatatoto-petiolaris Kondo—Australia
D. drummondii Lehm.—Australia = *D. barbigera*
D. elongata Exell. et Laundon—Angola
D. ericksonae Diels—Australia
D. erythrorhiza Lindley.—Australia
D. filicaulis Benth.—W. Australia = *D. menziesii*
D. filiformis Rafin—N. America
D. filiformis f. *tracyi* MacF.—Gulf States USA
D. filiformis Mazrimas = *filiformis* f. *filiformis* x *filiformis* f. *tracyi* = *D.* x 'california sunset'
D. filipes Turcs. = *D. huegelii*
D. fimbriata De Buhr—S.W. Australia
D. finlaysoni Wall. = *D. indica*

D. finlaysoniana Wall. ex Stein—Vietnam
D. flabellata Benth. = *D. platypoda*
D. flagellifera Col. = *D. binata*
D. flexicaulis Welw. ex Oliver—Tropical Africa = *D. affinis*
D. foliosa Hook. ex Planchon = *D. peltata* ssp. *foliosa*
D. fulva Planchon. = *D. petiolaris*
D. gigantea Lindley.—Australia
D. glabripes Harv.—S. Africa
D. glanduligera Lehm.—Australia
D. gracilis Hook. ex Planchon. = *D. peltata* ssp. *gracilis*
D. graminifolia St. Hil.—Brazil
D. grandiflora Bartl. = *D. pauciflora*
D. graniticola N. Marchant—S.W. Australia
D. hamiltonii C. Andrews—Australia
D. helianthemum Planchon. = *D. cistiflora*
heterophylla Lindley.—Australia
D. hexagyna Blanco = *D. indica*
D. hilaris Cham. et Schlechtd.—S. Africa
D. hirtella St. Hil.—Brazil = *D. montana* var. *hirtella*
D. huegelii Endl.—Australia
D. humbertii Exell et Laundon—Madagascar
D. humilis Planchon. = *D. stolonifera* ssp. *humilis*
D. x *hybrida* MacF. = *D. filiformis* x *intermedia*—N.J., USA
D. incisa A. Rich.—Cuba = *U. incisa*
D. indica L.—Asia, Tropical Australia, S. Africa
D. indica ssp. *robusta* Bailey—Australia
D. insolita Taton—Belgium, Congo
D. intermedia Hayne—Europe, N. America, Guyana
D. intricata Planchon. = *D. subhirtella*
D. iaieteurensis Brumm-Ding Guy
D. katangensis Taton—C. Africa
D. lanata Kondo—Australia
D. leucantha Shinners—S.E. States USA = *D. brevifolia*
D. leucoblasta Benth.—Australia
D. lingolata Col. = *D. arcturi*
D. linearis Goldie.—N. America
D. x *linglica* Nagamato = *D. linearis* x *anglica*
D. x *linpill* Nagamato = *D. anglica* x *capillaris*
D. x. *linthulata* Nagamato = *D. anglica* x *spatulata*
D. longifolia = *D. anglica*

Drosera capensis

D. loureirii Hook. et Arn. = *D. spathulata*
D. lovellae Bailey—Australia = *D. spathulata*
D. lunata Ham. = *D. peltata* var. *lunata*
D. macedonica Kolanin—Macedonia = *D. anglica*
D. macloviana Gandoger—Falkland Islands = *D. uniflora*
D. macrantha Endl.—Australia
D. macrantha ssp. *macrantha* Endl.
D. macrantha ssp. *planchonii* Hook. ex Planchon
D. macrophylla Lindley.—Australia
D. madagascariensis DC.—Madagascar, Tropical Africa
D. makinoi Masam. = *D. indica*
D. marchantii De Buhr—S.W. Australia
D. maritima St. Hill.—Brazil = *D. brevifolia*
D. menziesii R. BR.—Australia
D. menziesii ssp. *menziesii* Br. ex. Dc *D. menziesii* ssp. *mysanosepala* Diels, Marchant—Australia
D. metziana Gandoo—India = *D. indica*
D. micrantha Lehm = *D. paleacea*
D. microphylla Endl.—Australia
D. microphylla var. *macropetala* Diels
D. miniata Diels = *D. leucoblasta*
D. minutiflora Planchon = *D. paleacea*
D. minutula Col. = *D. pygmaea*
D. modesta Diels—Australia
D. montana St. Hil.—Brazil, Venezuela
D. montana var. *robusta* Diels—Venezuela
D. montana var. *roraimae* Klotzsch—Venezuela
D. myriantha Planchon.—Australia
D. x 'nagamoto' Nagamato = *D. anglica x spathulata*—Hort. (Japan)
D. natalensis Diels—S. Africa
D. neesii Lehm.—Australia
D. neesii ssp. *neesii* Lehm, Marchant—S.W. Australia
D. neesii ssp. *sulphurea* Lehm.
D. neo-caledonica Hamet—New Caledonia
D. nipponica Masamune = *D. peltata* ssp. *lunata*
D. nitidula Planchon.—Australia
D. x. obovata Koch = *D. rotundifolia x anglica*—Asia
D. oblanceolata Ruan—China
D. occidentalis Morr.—W. Australia
D. omissa Diels = *D. ericksonae*
D. paleacea DC.—Australia
D. pallida Lindley.—Australia
D. panamensis Lorrea & Taylor—Panama
D. parvifolia St. Hil.—Brazil = *D. montana*
D. parvula Planchon.—Australia
D. pauciflora Banks ex DC.—S. Africa
D. pauciflora var. *minor* Sond. in F.C. = *D. cistiflora*
D. pedata Pers. = *D. binata*
D. peltata SM. ex. Willd.—Australia, Japan, Formosa
D. peltata var. *lunata* Clarke—Japan, Formosa
D. penduhflora Planchon = *D. menziesii*
D. penicillaris Benth. = *D. menziesii*
D. petiolaris R. BR. ex DC.—Australia

D. pilosa Exell. et Laundon—Cameroons, Kenya, Tanganyika
D. planchonii Hook. F. ex Planchon. = *D. macrantha* ssp. *planchonii*
D. platypoda Turcz.—Australia
D. platystigma Lehm.—Australia
D. polyneura Col. = *D. arcturi*
D. praefolia Tepper = *D. whittakeri* ssp. *praefolia*
D. prolifera C.T. White—N.E. Australia
D. propinqua R. Cunn. = *D. spathulata*
D. pulchella Lehm.—Australia
D. purpurascens Schlott = *D. stolonifera* ssp. *stolonifera*
D. pusilla H.B.K.—Venezuela
D. pycnoblasta Diels—Australia
D. pygmaea DC.—Australia, New Zealand
D. radicans N. Marchant—W. Australia
D. ramellosa Lehm.—Australia
D. ramentacea Burch. ex. DC.—S. Africa
D. ramentacea var. *glabripes* Harv. ex Planchon. = *D. glabripes*
D. regia Step.—S. Africa
D. roraimae Maguire and Laundon—Venezuela
D. rosulata Lehm. = *D. bulbosa*
D. rotundifolia L.—N. hemisphere
D. rotundifolia x intermedia Callier—USA
D. ruahinensis Col. = *D. arcturi*
D. rubiginosa Heckel—New Caledonia = *D. neocalidonica*
D. sanariapoana Steyermark = *D. cayennensis*
D. schizandra Diels—Australia
D. scorpioides Planchon.—Australia
D. scorpioides var. *brevipes* Benth
D. serpens Planchon. = *D. indica*
D. sessilifolia St. Hil.—Brazil, Guyana
D. sewelliae Diels = *D. platystigma*
D. spathulata LaBill.—Australia, New Zealand, Japan
D. speciosa Presl = *D. cistiflora*
D. spiralis St. Hil.—Brazil = *D. graminifolia*
D. squamosa Benth. = *D. erythrorhiza*
D. stenopetala Hook.—New Zealand
D. stolonifera Endl.—Australia
D. stolonifera ssp. *compacta* Marchant
D. stolonifera ssp. *humilis* Planchon, Marchant
D. stolonifera ssp. *rupicola* Marchant
D. stolonifera ssp. *stolonifera* Endl
D. stricticaulis Diels. O. Sarg—Australia
D. stylosa Col. = *D. auriculata*
D. subtilis N. Marchant—N.W. Australia
D. subhirtella Planchon.—Australia
D. subhirtella ssp. *moorei* (Diels) Marchant—Australia
D. subhirtella ssp. *subhirtella* Planchon.—Australia
D. sulphurea Lehm. = *D. neesii*
D. tenella H.B.K.—Argentina = *D. pusillis*
D. tenella var. *esmeraldae* Steyermark = *D. capillaris*
D. thysanosepala Diels = *D. menziesii*

D. tracyi = *D. filiformis* f. *tracyi*
D. triflora Col. = *D. spathulata*
D. trineryia Spreng.—S. Africa
D. umbellata Lour.—China = *D. androsae umbellata*
D. uniflora Willd.—S. America
D. violacea Willd. = *D. cistiflora*
D. villosa St. Hil.—Brazil
D. whittakeri Planchon.—Australia
D. whittakeri ssp. *praefolia* (Tepper) Black—S. Australia
D. zeyheri Salter = *D. cistiflora*
D. zonaria Planch.—Australia

Drosophyllum

Family: DROSERACEAE

Drosophyllum lusitanicum Link—Spain

Genlisea

Family: LENTIBULARIACEAE

G. africana Oliver—Trop. Africa
G. africana ssp. *siapfii*—W. and C. Africa
G. anfractoosa Tutin = *G. filiformis*
G. angolensis Good—Angola, Zaire
G. aurea A. St. Hil.—Brazil
G. biloba Benj. = *G. violacea*
G. cylindrica Sylven = *G. violacea*
G. exmeraldae Steyerm. = *G. pygmaea*
G. filiformis A. St. Hil.—Northern South America
G. glabra P. Taylor—Venezuela
G. glandulosissima R.E. Fries—Zambia
G. guianenesis N.E. Br.—Guyana, Venezuela, Brazil
G. hispidula Stapf—Trop. and S. Africa
G. hispidula ssp. *subglabra*—E. Africa
G. luetzelburgii (Merl Ex Luetzelb.) Norman = *G. guianensis*
G. luteo-viridis Wright Apud Sauv. = *G. filiformis*
G. margaretae Hutchinson—Zambia, Mauagascak, Tanzania
G. minor A. St. Hil. = *G. aurea*
G. nigrocaulis Steyerm. = *G. pygmaea*
G. ornata Marl Ex. Benj. = *G. aurea*
G. oxycentron P. Taylor = *G. pygmaea*
G. pulchella Tutin = *G. repens*
G. pusilla Warm. = *G. repens*
G. pygmaea A. St. Hil.—Northern, South America
G. recurva Bosser = *G. margaretae*
G. relexa Benj. = *G. violacea*
G. repens Benj.—Brazil, Venezuela, Guyana, Paraguay
G. roraimensis N.E. Br.—Venezuela
G. sanariapoarra Steyerm.—Venezuela
G. stapfii A. Chevalier = *G. africana* ssp. *stapfii*
G. subgflabra Stapf = *G. hispidula* ssp. *subglabra*
G. subviridis Hutchinson = *G. africana* ssp. *africana*
G. uhoinate Taylor & Fromn Trinta-Brazil
G. violacea St. Hill.—Brazil

Heliamphora

Family: SARRACENIACEAE

H. heterodoxa Steyermark—Venezuela
H. heterodoxa var. *exappendiculata* Maguire & Steyermark—Chimantha Tepui-Ven
H. heterodoxa var. *glabra* Maguire
H. heterodoxa var. *heterodoxa* Steyermark—Mount Ptari-tepui, Venezuela
H. ionasi Maguire—Mt Ilu-tepui, Venezuela
H. macdonaldae Gleason—Mount Duida, Venezuela = *H. tatei* f. *macdonaldae*
H. minor Gleason—Mount Auyan-tepui, Venezuela
H. neblinae = *H. tatei* var. *neblinae*

Heliamphora minor

H. neblinae var. *neblinae* Maguire—Cerro de la Neblina
H. neblinae var. *parva* Maguire
H. neblinae var. *viridis* Maguire
H. nutans Bentham—Mount Roraima, Venezuela
H. tatei Benth—Venezuela
H. tatei var. *macdonaldae* Maguire
H. tatei var. *neblinae*
H. tatei var. *tatei* Gleason—Mount Duida, Mount Huachamacari, Venezuela
H. tyleri Gleason = *Tatei gleason*

IBICELLA
FAMILY: MARTYNIACEAE

I. lutea Lindl Van Eseltine—S. America
Martynia lutea = Ibicella lutea
Proboscidea lutea = Ibicella lutea

NEPENTHES
FAMILY: NEPENTHACEAE

N. x accentual koto Kawase = *thorelii x hookeriana*
N. x aichi Kondo & Sakai = *N. thorelii x balfouriana*
N. alata Blanco—Malaysia, Philippines, Sumatra
N. alba Ridl. = *N. gracillima*
N. albo-lineata Bail. = *N. mirabilis*
N. albo-marginata Lobb ex Lindl.—Malaysia, Sumatra, Borneo
N. alicae Bail. = *N. mirabilis*
N. x allardii Lynch = *N. veitchii x maxima*
N. x amabilis Nichols = *N. hookeriana x rafflesiana*
N. x ambrosial koto Kawase = *N. trichocarpa x hookeriana*
N. x amesiana Veitch. = *N. rafflesiana x hookeriana*
N. ampullaria Jack—Malaysia, New Guinea, Borneo, Sumatra
N. anamensis MacFarlane—E.S. Asia
N. angustifolia Mast. = *N. gracilis*
N. x arakawae Toyoshima = *N. mixta x alata*
N. armbrustae Bail. = *N. mirabilis*
N. x atropurpurea = *N. sanguinea x maxima 'superba'*
N. x. atro-sanguinea Mast. = *N. distillatoria x sedenii*
N. x balfouriana Mast. = *N. mixta x mastersiana*
N. x balmy koto Kawase = *N. thorelii x maxima*
N. beccariana MacF. = *N. mirabilis*
N. bellii Kondo—Philippines
N. bernaysii Bail. = *N. mirabilis*
N. blancoi Bl. = *N. alata*
N. bicalcarata Hook—Borneo

N. x bohnickii Bonstedt = *N. (mixta x maxima) x (mixta x maxima)*
N. x boissiana Desloges = *N. tiveyi x morganiana*
N. x boissiense Lecoufle = *N. gracilis x superba*
N. bongso Korth.—Sumatra
N. boschiana Korth.—Borneo
N. brachycarpa Merr. = *N. philippinensis*
N. burbidgeae Hook. F. ex Burbidge—Borneo
N. burkei Masters—Philippines
N. campanulau Kurata—Borneo
N. x caroli-schmidtii Bonstedt = *N. mixta x allardii*
N. carunculata Danser—Sumatra
N. celebica Hook = *N. maxima*
N. x chelsoni Veitch. ex. Masters = *N. dominii x hookeriana*
N. x chelsonii excellens Veitch = *N. rafflesiana x chelsonii*
N. cholmondeleyi Bail. = *N. mirabilis*
N. cincta Masters = *N. albo-marginata x northiana*
N. clipeata Danser—Borneo
N. x coccinea Mast. = *hookeriana x mirabilis*
N. x compacta Baines = *hookeriana x mirablis*
N. copelandii MacF. = *N. alata x courtii* Veitch.
= *N. gracilis x dominii x curtisii* Masters.
= *N. maxima*
N. x cylindrica Veitch. = *N. distillatoria x veitchii*
N. deaniana MacF.—Philippines
N. decurrens MacF.—Borneo
N. deslogesii Desloges = *N. tiveyi x mixta*
N. densiflora Danser—Sumatra
N. dentata Kurata—Sulawesi = *N. hamata*
N. x dicksoniana Lindsay = *N. rafflesiana x veitchii*
N. x Director George T. Moore Pring
N. distillatoria L.—Sri Lanka
N. x dominii Veitch. ex Masters = *N. rafflesiana x gracilis*
N. x Dr. D.C. Fairburn Pring
N. x Dr. Edgar Anderson Pring = *N. chelsonii x dominii*
N. x dormanniana B.S. Wms = *N. mirabilis x sedenii*
N. x dreamy koto Kawase = *N. thorelii x veitchii*
N. dubia Danser—Sumatra = *N. bongo x inermis*
N. x dyeriana MacF. = *N. mixta x dicksoniana*
N. echinostoma Hook. = *N. mirabilis*
N. x ecstatic koto Kawase = *N. thorelii x maxima*
N. x edinensis Lindsay = *N. rafflesiana x chelsoni*
N. edwardsiana Low ex Hook.—Borneo = *N. villosa*
N. x effulgent koto Kawase = *N. mirabilis x thorelii*
N. ephippiata Danser—Borneo
N. eustachys Miq. = *N. alata*
N. x excelsa = *N. veitchii x sanguinea*
N. x excelsior B.S.Wms. = *N. rafflesiana x hookeriana*
N. x eyermanni Sieb = *N. mirabilis x hookeriana*
N. eymai (Kurata) Turnbull & Middleton = *N. infundibuiformis*
N. fallax Beck. = *N. stenophylla*
N. x formosa Hort & Kew = *N. chelsonii x distillatorias*
N. x fournieri Gautier = *N. northiana x mixta*

N. x fukakusana Toyoshima = *N. rafflesiana x dyeriana*
N. x fulgent koto Kawase = *N. thorelii x fusca*
N. fusca Danser—Borneo
N. x fushimiensis Toyoshima = *N. globamphora x thorelii*
N. x F.W. Moore Veitch = *N. mixta x dicksoniana*
N. x gamerii Desloges = *N. tiveyi x mixta*
N. garrawayae Bail. = *N. mirabilis*
N. gautieri Gautier = *N. northiana x mixta*
N. geoffrayi Lecomte—S.E. Asia
N. x Gerald Ulrici Pring = *N. chelsonii x dominii*
N. glabrata Turnbull & Middleton—Sulawesi
N. globamphora Toyoshima & Kurata = *N. bellii*
N. x goebelii Bonstedt = *N. mixta x maxima*
N. x goettingensis Bonstedt = *N. mixta x dicksonianas*
N. graciliflora Elm. = *N. alata*
N. gracilis Korth.—Borneo, Malaysia, Singapore, Sulawesi, Sumatra
N. gracillima Ridley—Malaysia
N. gracillima var. *major* Rid.
N. x grandis Desloges = *N. maxima* 'superba' x *northiana* 'pulchra'
N. gymnamphora Miq.—Sumatra, Borneo, Java
N. x hachijo Okuyama = *N. mirabilis complex x mirabilis*
N. hamata Turnbull & Middleton—Sulawesi
N. harryana Burb. = *N. villosa* ssp. *hybrid villosa x edwardsiana*
N. hemsleyana MacF. = *N. rafflesiana*
N. x Henry Shaw Pring = *N. chelsonii x dominii*
N. x henryana Nicholson = *N. sedeni x hookeriana*
N. x hibberdii Nichols = *N. hookeriana x sedeni*
N. hirsuta Hook.—Borneo
N. hispida Beck. = *N. hirsuta*
N. x hoelscheri Bonstedt = *N. mixta x rufescens*
N. x hookerae Beck = *N. rafflesiana x mirabilis*
N. x hookeriana Low = *N. rafflesiana x ampullaria*—Borneo, Sumatra, Malaysia
N. x hybrida Veitch & Masters = *N. khasiana x gracilis*
N. x hybrida maculata Veitch et Dombrain = *N. hybrida*
N. x hybrida maculata elongata Hort ex Burbidge = *gracilis x dominii*
N. x Ille de France Y. Vezier = *N. lecouflei x mixta*
N. inermis Danser—Sumatra
N. infundibuliformis Turnbull & Middleton—Sulawesi
N. x intermedia Hort. ex Veitch. = *N. gracilis x rafflesiana*
N. insignis Danser—New Guinea
N. jardinei Bail. = *N. mirabilis*
N. x Joseph Cutak Pring = *N. chelsonii x dominiis*
N. x jumghuhnii Ridl. = *N. sanguinea x singalanaa*
N. kampotiana Lecomte—S.E. Asia
N. x Katherine Moore Pring = *N. chelsonii x dominii*
N. kennedyana F. Meull. = *N. mirabilis*
N. khasiana Hook.—Assam, India
N. x kikuchiae Okuyama = *N. oiso x maxima*
N. x kinabaluensis Kurata = *N. rajah x villosa*

N. klossii Ridley—New Guinea
N. x krausii Bonstedt = *N. mixta x allardii*
N. x ladenburgii Bonstedt = *N. mixta x maxima*
N. laevis Lindl. = *N. gracilis*
N. lanata Mast. = *N. veitchii*
N. x lawrenceana Mast. = *N. mirabilis x hookeriana*
N. x lecouflei Kusakabe = *N. mirabilis x thorelii*—Cambodia
N. leptochila Danser—Borneo
N. x Lieut. R. Bradford Pring = *N. chelsonii x dominii*
N. loddigesii Baxt. = *N. hookeriana*
N. x longicaudata Desloges = *N. maxima* 'superba' x *northiana* 'pulchra'
N. lowii Hook.—Borneo
N. x lyrata Veitch = *N. hybrida x rafflesiana*
N. x Dr. J. Macfarlane Veitch = *N. sanguinea x maxima* 'superba'
N. macfarlanei Hemsl.—Malaysia
N. macrovulgaris Turnbull & Middleton—Borneo
N. madagascariensis Poir.—Madagascar
N. x Maria-Louisa Gautier = *N. northiana x mixta*
N. x masahiroi Kondo = *N. albo-marginata x thorelii*
N. x masamiae Toyoshima = *N. thorelii x maxima*
N. masoalensis Schmid-Hollinger—Madagascar
N. x mastersiana Veitch = *N. sanguinea x khasiana*
N. maxima Reinw.—Celebes, Borneo, New Guinea
N. melamphora BL. = *N. gymnamphora*
N. x mercieri Gautier = *N. northiana x maxima*
N. merrilliana Macf.—Philippines
N. x merrilliata = *N. merrilliana x alata* (natural hybrid)
N. x minamiensis Naito = *N. oiso x wrigleyana*
N. mirabilis Druce—SE Asia, Australia, New Guinea
N. x mixta Veitch. = *N. northiana x maxima*
N. x mizuho Kondo = *N. rafflesiana x dyeriana*
N. mollis Danser—Borneo
N. montrouzierii Dub. = *N. vieillardi*
N. moorei Bail. = *N. mirabilis*
N. x morganiana Veitch = *N. mirabilis x hookeriana*
N. muluensis Hotta—Borneo
N. x nagoya Kondo = *N. mixta x thorelii*
N. x nasu Okuyama = *N. thorelii x wrigleyana*
N. neglecta Macf.—Borneo
N. x Nel Horner Pring = *N. chelsonii x dominii*
N. neo-guineensis Macf.—New Guinea
N. x neufvilliana Bonstedt = *N. mixta x maxima*
N. x. nobilis Veitch = *N. sanguinea x maxima* 'superba'
N. northiana Hook.—Borneo
N. oblanceolata Ridley = *N. maxima*
N. x oiso Ikeda = *N. mixta sanderiana x maxima* 'superba'
N. x outramiana B.S. Wms. = *N. sedenii x hookeriana*
N. paniculata Danser—New Guinea
N. papuana Danser—New Guinea
N. x paradisae (Beck) Taplin = *N. mirabilis x hookeriana*
N. pascoensis Bail. = *N. mirabilis*
N. x patersonii Sieb & Wadley = *N. mirabilis x hookeriana*

N. x *paullii* Desloges = *N. tiveyi* x *mixta*
N. pectinata Danser—Sumatra
N. pervillei Bl.—Seychelles
N. x *petersii* Bonstedt = *N. mixta* x *Allardii*
N. petiolata Danser—Philippines
N. philippinensis Macf.—Philippines = *N. alata*
N. phyllamphora Stapf. = *N. alata* or *N. mirabilis*
N. x *picturata* Veitch = *N. mixta* x *dicksoniana*
N. pilosa Danser—Borneo
N. x *pitcheri* Mast. ex Millers = *N. paradisae* x *henryana*
N. x *princeps* Kondo = *N. mixta* x *dyeriana*
N. x *prosperity* Kubo = *N. mirabilis complex* x *dyeriana*
N. pumila Griff. = *N. sanguinea*
N. ramispina Ridl. = *N. gracillima*
N. x *ratcliffiana* Court = *N. mirabilis* x *hookeriana*
N. rafflesiana Jack—Malaysia, Borneo, Sumatra
N. x *rafflesiana pallida* Burbidge = *N. hybrida* x *rafflesiana*
N. rajah Hook—Borneo
N. reinwardtiana Miq.—Sumatra, Borneo, Malaysia
N. x *remilliensis* Macf. = *N. mixta* x *tiveyi*
N. x *reutheri* Bonstedt = *N. mixta* x *masteriana*
N. rhombicaulis Kurata—Sumatra
N. x *robusta* Mast. = *N. mirabilis* x *hookeriana*
N. x *roedigeri* Bonstedt = *N. mixta* x *maxima*
N. x *rokko* Yamakawa = *N. thorelii* x *maxima*
N. rowanae Bail. = *N. mirabilis*
N. x *rubro-maculata* Veitch = *N. hybrida maculata* x spp. incert. *borneensis*
N. rubro-maculata (Kurata) Turnbull & Middleton = *N. glabrata*
N. x *rufescens* Mast. = *N. distillatoria* x *courtii*
N. x *rutzii* Bonstedt = *N. mixta* x *tiveyi*
N. sanguinea Lindl.—Malaysia
N. x *sedenii* Veitch. ex Masters. = *N. gracilis* x *khasiana*
N. x *shinjuku* Fukaba ex Makino = *N. mixta* x *wrigleyana*
N. x *shioji* Kondo = *N. mixta* x *dyeriana*
N. x *siebrechitiana* Miller = *N. mirabilis* x *sedenii*
N. x *siebertii* Bonstedt = *N. mixta* x *Allardii*
N. x *simonii* Gautier = *N. northiana* x *mixta*
N. singalana Becc.—Sumatra
N. x *Sir W.T.T. Dyer* (Veitch) Macf. = *N.* x *dyeriana*
N. smilesii Hemsl. = *N. mirabilis*
N. spathulata Danser—Sumatra
N. spectabilis Danser—Sumatra
N. x *sprendida* Pitcher & Manda = *N. mirabilis* x *hookeriana*
N. spuria Beck = *N. northiana*
N. x *St Louis* Pring = *N. chelsoni* x *dominii*
N. x *stammieri* Bonstedt = *N.* (*mixta* x *maxima*) x (*mixta* x *maxima*)
N. stenophylla Mast.—Borneo, Malaysia
N. x *stewartii* Moore = *N. mirabilis* x *hookeriana*
N. surigaoensis Elm. = *N. merrilliana*

N. x *superba* B.S.Wms. = *N. sedenii* x *hookeriana*
N. x *suzue* Kondo = *N. mixta* x *thorelii*
N. x *takayuki sakai* Kondo & Sakai = *N. tobaica* x *thoreliis*
N. tentaculata Hook—Borneo, Celebes
N. teysmanniana Miq. = *N. albo-marginata*
N. thorelii Lecomte—Cambodia
N. x *tiveyi* Mast. = *N. maxima superba* x *veitchii*
N. tobaica Danser—Sumatra = *N. reinwardtiana*
N. x *tokuyoshi kondo* Kondo & Sakai = *St Louis* x *rafflesiana*
N. tomoriana Danser—Celebes
N. x *toyoshimae* Toyoshima = *N. truncata* x *thorellii*
N. x *trusmadiensis* J. Marabini = *N. edwardsiana* x *lowii*
N. treubiana Warb.—New Guinea, Sumatra
N. x *trichocarpa* Miq. = *N. ampullaria* x *gracilis* natural hybrid—Sumatra, Borneo, Malaysia
N. truncata Mac.—Philippines
N. x *truncata* x *petiolata* Kurata
N. x *tsujimoto* Takarazuka B.G. = *N. mastersiana* x *wrigleyana*
N. tubulosa MacF. = *N. mirabilis*
N. x *vallierae* Desloges = *N. triveyi* x *mixta*
N. veitchii Hook.—Borneo
N. x *ventrata* Sivertsen = *N. ventricosa* x *alata*
N. ventricosa Blanco—Philippines
N. vieillardi Hook.—New Caledonia
N. x *Ville de Rouen* Y. Vezier = *N. superba* x *mastersiana*
N. villosa Hook.—Borneo
N. x *williamsii* Mast. = *N. sedeni* x *hookeriana*
N. x *wittei* Veitch = *N. maxima* x *stenophylla*
N. x *wrigleyana* Veitch. ex Masters = *N. mirabilis* x *hookeriana*
N. x *yatomi* Kondo & Sakai = *N. thorelii* x *veitchii*
N. zeylanica Rafin. = *N. distillatoria*

Pinguicula

Family: Lentibulariaceae

P. acuminata Benth.—Mexico
P. acutifolia Michx. = *P. villosa*
P. agnata Casper—Mexico
P. alba Kuchl = *P. alpina*
P. albanica Griseb. = *P. hirtiflora*
P. albida Wright ex. Griseb.—Cuba
P. algida Malyschev—Russia
P. alpestris Persoon = *P. alpina*
P. alpina L.—Europe, Asia
P. antarctica Vahl—Chile, Argentina, Peru
P. arctica Eastwood = *P. vulgaris*
P. arveti Genty Apud Morot = *P. leptoceras*
P. australis Nutt. = *P. pumila*

P. bakeriana Sander = *P. moranensis*
P. bakeriana Sprague = *P. macrophylla*
P. balcanica Casper—Bulgaria, Yugoslavia, Albania, Greece
P. balcanica var. *tenuilaciniata* Casper—Greece
P. benedicta Barnh.—Cuba
P. bicolor Wolosczak = *P. vulgaris* f. *bicolor*
P. bohemica Krajina Ernst—Czechoslovakia
P. brachyloba Ledebour = *P. alpina*
P. caerulea Walt.—NC, SC, Ga, Fla
P. calyptrata H.B.K.—Colombia, Ecuador
P. campanulata Lam. = *P. lutea*
P. casabitoana Jiminez—Cuba
P. caudata Hemsley = *P. macrophylla* or *P. oblongiloba*
P. caudata Schlecht. = *P. moranensis*
P. caudata x gypsicola Stapf Horticultural hybrid
P. chilensis Clos—Chile, Argentina
P. cladophila Ernst = *P. casabitoana*
P. clivorum Standley et Steyerm.—Name doubtful
P. colimensis MC Vaugh et Mickel—Mexico
P. corsica Bern. et Gren. ex Gren. et Godr.—Corsica
P. crenatiloba D.C.—Mexico, Guatemala, Honduras, El Salvador, Panama
P. crystallina Sibth. ex Sibth. FT Smith—Cyprus
P. cyclosecta Casper—Mexico
P. davurica Link = *P. macroceras*
P. diversifolia Cuatr. = *P. elongata*
P. edentula Hook. = *P. lutea*
P. ehlersae spetae Fuchs.—Mexico
P. elatior Michx. = *P. caerulea*
P. eliae Sennen = *P. grandiflora*
P. elongata Benj.—Venezuela, Colombia
P. esseriana Kirchner—Mexico
P. filifolia Wright ex Griseb.—Cuba
P. flavescens Floerke = *P. alpina*
P. floridensis Chapm. = *P. pumila*
P. flos-mulionis Morr. = *P. moranensis*
P. gelida Schur = *P. alpina*
P. grandiflora Lam.—Ireland, Spain, France, Switz.
P. grandiflora f. *pallida* (Gaudin Casper)—France
P. grandiflora ssp. *rosea* (Mutel Casper)—France
P. grandulosa Trautv. et Meyer = *P. variegata*
P. gypsicola Brandegee—Mexico
P. gypsophila Wallroth = *P. vulgaris*
P. hellwegeri Murr = *P. leptoceras*
P. heterophylla Benth.—Mexico
P. hirtiflora Ten.—Italy, East Mediterranean
P. hirtiflora f. *pallida* Casper—Italy, E. Med.
P. hirtiflora var. *louisii* (Markgraf) Ernst—Italy, E. Med.
P. hirtiflora var. *megaspitaea* (Boiss. et Heldr.) Schindler—Italy, E. Med.
P. huilensis Cuatr. = *P. calyptrata*
P. x hybrida Wettst. = *P. alpina x vulgaris*
P. hyperborea Grandoger = *P. alpina*

P. imitatrix Casper—Mexico
P. inaequilobata Sennen = *P. grandiflora*
P. involucrata D.C. = *P. villosa*
P. involuta Ruiz et Pav.—Bolivia, Peru
P. ionantha Godfrey—Florida
P. jackii Barnh.—Cuba
P. jackii var. *parviflora* Ernst—Cuba
P. juratensis Bernard = *P. grandiflora* f. *pallida*
P. kamischatica Roemer et Schultes = *P. macroceras*
P. kondoi Casper—Mexico
P. laeta Pantocsek = *P. hirtiflora*
P. lateciliata MC Vaugh et Mickel = *P. cyclosecta*
P. leptoceras Reichb.—Switz., Austria, Italy, France
P. lignicola Barnh.—Cuba
P. lilacin Schlecht. et Cham.—Mexico
P. longifolia Ram. ex D.C.—Spain, France, Italy
P. longifolia ssp. *caussensis* Casper—France
P. longifolia ssp. *longifolia*—France
P. longifolia ssp. *reichenbachiana* (Schindler) Casper—France, Italy
P. louisii Markgraf = *P. hirtiflora* var. *louisii*
P. lusitanica L.—Portugal, France, Great Britain, N. Africa, Spain
P. lutea Walt.—La, Miss, Ala, NC, SC, Ga, Fla
P. macroceras Link—Japan, USSR, N.W. North America
P. macrophylla H.B.K.—Mexico
P. macrostyla Benj. = *P. involuta*
P. magellanica Comm. ex. Franchet = *P. antarctica*
P. megaspilaea Boiss. et Heldr. = *P. hirtiflora* var. *megaspilaea*
P. merinoana Sennen = *P. grandiflora*
P. microceras Cham. = *P. macroceras*
P. moranensis H.B.K.—Mexico, Guatemala, El Salvador
P. nana Mart. et Gal. = *P. crenatiloba*
P. nevadensis (Lindbg.) Casper—Spain
P. norica Beck = *P. vulgaris*
P. oblongiloba C.D.—Mexico
P. obtusa Herb. Banks Descr. Benj. = *P. antarctica*
P. obtusiloba D.C. = *P. lilacina*
P. obtusiloba Ernst = *P. agnata*
P. occyptera Rchb. ex Benj.—name doubtful
P. pallida Turcz. = *P. alpina*
P. parvifolia Robinson—Mexico
P. planifolia Chapm.—Fla, Miss.
P. primuliflora Wood et Godfrey—Ala, Fla, Miss
P. pumila Michx.—Tex, La, Ala, NC, Ga, Fla, Bahamas
P. purpurea Willd. = *P. alpina*
P. ramosa Miyoshi ex Yatabe—Japan
P. reichenbachiana Schindler = *P. longifolia* ssp. *reichenbachiana*
P. reuteri Genty = *P. grandiflora* f. *pallida*
P. reuteri Schindler = *P. grandiflora* ssp. *rosea*
P. rotundifolia Studnicka—Mexico
P. rosei Watson = *P. moranensis*

P. scopulorum T.S. Brandegee = *P. lilacina*
P. x scullyi Druce = *P. grandiflora x vulgaris*—Ireland, France
P. sharpii Casper & Kondo—Mexico
P. sibirica Vest—name doubtful
P. sodalium Fourn. = *P. moranensis*
P. spathulata Ledebour—name doubtful
P. subaequalis Stokes = *P. lusitanica*
P. vallisneriifolia Webb—Spain
P. vallisneriifolia lus. brevifolia Casper—Spain
P. variegata Turcz.—Siberia
P. villosa L.—Alaska, Canada, Sweden, Norway, Finland, USSR
P. villosa f. *albiflora* Frodin—N. America
P. villosa lus. ramosa Casper—N. America
P. violacea Michx. = *P. pumila*
P. vulgaris L.—Europe, Siberia, N. America
P. vulgaris f. *albida* (Behm. Neumann)—N. America
P. vulgaris f. *bicolor* (Nordst. ex Fries) Neumann—N. America
P. zecheri Speta & Fuchs.—Mexico

Sarracenia

Family: SARRACENIACEAE

S. adunca = *S. minor*
S. x ahlesii Bell et Case = *S. alata x rubra*—Ala, Miss
S. alata Wood—Ala, Miss. La, Tex.
S. alata x psittacina—Ala, Miss
S. x areolata MacFar. = *S. alata x leucophylla*—Ala, Miss
S. x cantabridgiensis = *S. x excellens*
S. x catesbaei Elliott = *S. flava x purpurea*—Va, NC, SC, Ga, Fla, Ala
S. x chelsonii Masters = *S. purpurea x rubra*—NC, SC, Ga, Fla, Ala, Miss
S. x courtii Masters = *S. psittacina x purpurea*—Ga, Fla, Ala, Miss
S. x diesneriana Hefka = *S. courtii x flava*—HORT.
S. drummondii Croom = *S. leucophylla*
S. x evedine = *S. catesbaei x leucophylla*
S. x excellens Nichols = *S. leucophylla x minor*—Ga, Fla
S. x exornata S.G. = *S. alata x purpurea*—Ala, Miss
S. x farnhamii = *S. x readii*
S. flava L.—Va, NC, SC, Ga, Fla, Ala
S. flava x psittacina—Ga, Fla, Ala
S. x formosa Veitch ex Masters = *S. minor x psittacina*—Ga, Fla
S. x gilpini Bell et Case = *S. psittacina x rubra*—Ga, Fla, Ala, Miss
S. x harperi Bell = *S. flava x minor*—NC, SC, Ga, Fla
S. x illustrata Nichols. = *S. alata x catesbaei*—Hort.
S. x laschkei Hefka = *S. courtii x mooreana*—Hort.
S. leucophylla Raf.—Ga, Fla, Ala, Miss
S. x Marston Mill = *S. leucophylla x catesbaei x flava*
S. x melanorhoda Hort. Veitch = *S. catesbaei x purpurea*—Hort.
S. minor Wat.—NC, SC, Ga, Fla
S. x mitchelliana Nichols. = *S. leucophylla x purpurea*—Ga, Fla, Ala, Miss
S. x mooreana Veitch = *flava x leucophylla*—Ga, Fla, Ala
S. oreophila (Kearney) Wherry—Ga, Ala
S. x popei Masters = *S. flava x rubra*—NC, SC, Ga, Fla, Ala
S. psittacina Michx.—Ga, Fla, Ala, Miss, La
S. purpurea L.—Eastern North America
S. purpurea gibbosa = *S. purpurea* ssp. *purpurea*
S. purpurea ssp. *purpurea* f. *heterophylla* (Eaton) Fernald—Newfoundland, Nova Scotia, Mass, NJ, Michigan, Ontario, Northern Range
S. purpurea ssp. *purpurea* Wherry—Northern Range to NJ
S. purpurea ssp. *venosa* Raf.—Southern range from NJ
S. x readii Bell = *leucophylla x rubra*—Ga, Fla, Ala, Miss
S. x rehderi Bell = *minor x rubra*—NC. SC, Ga, Fla

Sarracenia leucophylla

S. rubra Walt.—NC, SC, Ga, Fla, Ala, Miss
S. rubra ssp. *alabamensis* (Case & Case) Schnell—Ala
S. rubra ssp. *gulfensis* Schnell—Fla
S. rubra ssp. *jonesii* (Wherry) Wherry—NC, SC
S. rubra ssp. *wherryi* (Case & Case) Schnell—Ala
S. x sanderiana Hort. Sanders = *S. leucophylla x readii*—Hort.
S. sledgei MacFar. = *S. alata*
S. x swaniana WM. Robinson = *S. minor x purpurea*—NC, SC, Ga, Fla
S. x umlauftiana Hefka = *S. courtii x wrigleyana*—Hort.
S. undulata = *S. leucophylla*
S. variolaris = *S. minor*
S. x vetteriana Hefka = *S. illustrata x catesbaei*—Hort.
S. x vittata maculata Nichols. = *S. purpurea x chelsonii*—Hort.
S. x vogeliana Hefka = *S. courtii x catesbaei*—Hort.
S. x willisii Hort. Veitch = *S. courtii x melanorhoda*—Hort.
S. x willmottae Hort. Bruce = *S. melanorhoda*
S. x wrigleyana S.G. = *S. leucophylla x psittacina*—Ga, Fla, Ala, Miss

Triphyophyllum

Family: DIONCOPHYLLACEAE

Ouratea glomerata = *Triphyophyllum peltatum*
Triphyophyllum peltatum Baill.—W. Equatorial Africa, Sierra Leone, Liberia.

Utricularia

Family: LENTIBULARIACEAE

U. aberrans = *U. welwitschii*
U. adenantha = *U. amethystina* s.g. = *U. hirtella*
U. adpressa Salzm. ex St Hil. et Gir.—Northern South America
U. affinis = *U. uliginosa* s.g.
U. afromontana = *U. livida*
U. alata = Prob. *U. bifida*
U. alba Hoffmg. ex Link = *U. obtusa*
U. albiflora R. Br.—Australia
U. albina = *U. caerulea*
U. albo-caerulea Dalz—India
U. alpina Jacquin—S. and C. America, West Indies
U. alutacea = *U. amethystina*
U. amazonasana = *U. hydrocarpa*
U. amethystina St. Hil. et Gir.—Florida to Brazil
U. amoena = *U. myriocista*

U. andicola = *U. livida*
U. andongensis Hiern—Guinea to Zambia and Angola
U. angolensis = *U. subulata*
U. angulosa = *U. juncea*
U. angustifolia = *U. hispida*
U. anomala = *U. obtusa*
U. antennifera P. Taylor—W. Australia
U. anthropophora = *U. striatula*
U. aphylla = *U. obtusa*
U. appendiculata A.E. Bruce—Cameroon to Mozambique and Madagascar
U. arenaria A. DC.—Senegal to Ethiopia to S.W. and S. Africa, Madagascar, India
U. arcuata Wight—India
U. arnhemica P. Taylor—N. Australia
U. arenicola = *U. nana*
U. arrojadensis = *U. longifolia*
U. asplundii Taylor—Ecuador, Columbia
U. aurea Lour.—India to Australia, Japan
U. aureola = *U. adpressa*
U. aureolimba = *U. adpressa*
U. aureomaculata Steyerm.—Venezuela
U. australis R. Br.—Trop. and S. Africa, Europe, temp. Asia to Japan, Australia, New Zealand
U. ayacuchae = *U. cucullata*
U. baldwinii = *U. hispida*
U. bangweolensis = *U. reflexa* var. *reflexa*
U. baouleensis A. Chev. = *U. foveolata*
U. barbata = *U. chrysantha*
U. baueri = *U. caerulea*
U. baumii = *U. spiralis* var. *spiralis*
U. benjaminiana Oliver—Guyana, Suriname, trop. Africa, Madagascar, Trinidad
U. benthamii Taylor—Australia
U. bicolor = *U. amethystina*
U. bifida L.—India to Australia, Japan
U. bifidocalcar = *U. obtusa*
U. biflora Lam. = *U. gibba*
U. billardieri = *U. dichotoma*
U. biloba R. Br.—S.E. Asia, Australia, New Guinea
U. biovularioides Taylor—Brazil
U. bipartita = *U. biflora*
U. bisquamata Schrank—Madag., S. Africa
U. blanchetti A. DC.—Brazil
U. bolivarana = *U. amethystina*
U. bosminifera Ostenfeld—Thailand
U. botecudorum = *U. foliosa*
U. brachiata Wight—India
U. brachyceras = *U. capensis*
U. bracteata Good—Angola, Zambia, Zaire
U. bradei = *U. subulata*
U. bramadensis = *U. longifolia*
U. bremii Heer—Europe
U. brevicornis = *U. ochroleuca*

U. brevilabris = *U. minutissima*
U. breviscapa Wright ex Griseb.—Cuba, Guyana, Brazil
U. bryophila = *U. mannii*
U. buntingiana Taylor—Venezuela
U. butanensis = *U. foliosa*
U. caerulea L.—India to Australia, Japan
U. caerulea var. *stricticaulis* = *U. stricticaulis* Koenig Stapf ex Gamble
U. calliphysa Stapf = *U. minutissima*
U. calumpitensis = *U. aurea*
U. calycifida Benj.—Guyana, Venezuela, Suriname
U. campbelliana Oliver—Venezuela, Guyana
U. canacorum Pellegr. = *U. novae-zelandiae*
U. capensis Spreng. = *U. bisquamata*
U. capillaris = *U. subulata*
U. cappilliflora F. Muell.—Australia
U. cavalerii = *U. caerulea*
U. cearana Steyerm. = *U. adpressa*
U. cecilii Taylor—India
U. ceratophylla Michx. = *U. inflata*
U. ceratophylloides = Prob. *U. biloba*
U. cervicornuta = *U. benjaminiana*
U. charnleyensis = *U. caerulea*
U. charoidea = *U. reflexa* var. *reflexa*
U. cheiranthos Taylor—N. Australia
U. chiribiquetensis A. Fernandez—Colombia, Venezuela
U. choristotheca Taylor—Suriname
U. christopheri Taylor—Himalaya
U. chrysantha R. Br.—Australia, New Guinea
U. circumvoluta Taylor—N. Australia
U. clandestina = *U. geminiscapa*
U. cleistogama = *U. subulata*
U. coccinea = *U. hydrocarpa*
U. colensoi = *U. lateriflora*
U. colorata = *U. laxa*
U. concinna = *U. jamesoniana*
U. conferta = *U. exoleta*
U. congesta = *U. simulans*
U. connellii = *U. pubescens*
U. corenophora Taylor—Burma, Thailand
U. cornuta Michx.—N. America, Bahamas, Cuba
U. costata Taylor—Venezuela
U. crenata Vahl = *U. obtusa*
U. cucullata St. Hil. et Gir.—S. America
U. cuspidata = *U. calycifida*
U. cutleri = *U. viscosa*
U. cyanea R. Br. = *U. uliginosa*
U. cymbantha Oliver—South Africa
U. damazioi = *U. amethystina*
U. dawsoni = *U. amethystina*
U. deightonii = *U. pubescens*
U. delicata = *U. capensis*
U. delicatula Cheesm.—Australia, New Zealand
U. delphinoides Thorel ex Pellegr.—S.E. Asia

U. denticulata = *U. livida*
U. determannii Taylor—S. America
U. dichotoma Labill.—Australia
U. dimorphantha Makino—Japan
U. diploglossa = *U. reflexa* var. *reflexa*
U. dissectifolia = *U. flaccida*
U. dregei = *U. livida*
U. dunlopii Taylor—Australia
U. dunstaniae F.E. Lloyd—Australia
U. dusenii = *U. nephrophylla*
U. eburnea = *U. livida*
U. ecklonii = *U. capensis*
U. elephas = *U. cucullata*
U. elevata = *U. livida*
U. elmeri = *U. heterosepala*
U. emarginata = *U. obtusa*
U. endresii Reichb.—C. America
U. engleri = *U. livida*
U. equiseticaulis = *U. graminifolia*
U. erectiflora St. Hil. et Gir.—Northern South America
U. evrardii Pellegr. = *U. minutissima*
U. exigua = *U. subulata*
U. exilis = *U. arenaria*
U. exoleta R. Br. = *U. gibba*
U. falcata = *U. spiralis* var. *tortilis*
U. fasciculata Roxb. = *U. aurea*
U. fernaldiana = *U. pubescens*
U. fibrosa Walt = *U. gibba*
U. filiformis = *U. subulata*
U. fimbriata Kunth—Colombia, Venezuela
U. firmula Welw. ex Oliver—Africa, Madagascar
U. fistulosa Taylor—W. Australia
U. flaccida A. DC.—S. America
U. flexuosa = *U. aurea*
U. floridana Nash—Florida, Georgia, N. and S. Carolina
U. fluitans = *U. aurea*
U. fockeana = *U. hydrocarpa*
U. foliosa L.—Florida to Argentina, Trop. Africa, Madagascar, Galapagos Islands
U. fontana = *U. tricolor*
U. forgetiana = *U. longifolia*
U. fornicata Le Conte = *U. biflora*
U. forrestii Taylor—China
U. foveolata Edgew—Africa, S.E. Asia, Australia, China, India, Madagascar
U. fulva F. Muell.—Australia
U. furcata Pers. = *U. obtusa*
U. furcellata Oliver—India
U. fusiformis = *U. tricolor*
U. galloprovincialis = *U. australis*
U. garratii Taylor—Thailand
U. gayana = *U. obtusa*
U. geminiloba Benj.—Brazil
U. geminiscapa Benj.—N.E. USA

U. genliseoides = *U. hirtella*
U. gentryi = *U. livida*
U. geoffrayi Pellegr.—S.E. Asia, Thailand
U. georgei Taylor—W. Australia, Africa
U. gibba L.—USA & Japan
U. gibba f. *natans* Komiya—Japan
U. gibbsiae = *U. scandens* ssp. *scandens*
U. giletii = *U. benjaminiana*
U. glazioviana = *U. neottioides*
U. globulariifolia = *U. tricolor*
U. glueckii = *U. hispida*
U. goebelii = *U. purporeo-caerulea*
U. gomezii = *U. tricolor*
U. graminifolia Vahl—India to New Guinea
U. grandivesiculosa = *U. reflexa* var. *reflexa*
U. graniticola = *U. pubescens*
U. griffithii = *U. uliginosa*
U. guianensis Splitg. ex De Vries = *U. foliosa*
U. guyanensis A. DC.—Trinidad, Honduras, Guyana
U. gyrans = *U. spiralis* var. *tortilis*
U. hamiltonii F.E. Lloyd—Australia
U. harlandi Oliver = *U. striatula*
U. helix Taylor—W. Australia
U. herzogii = *U. neottioides*
U. heterochroma Steyerm.—Venezuela
U. heterosepala Benj.—Philippines
U. hians = *U. prehensilis*
U. hintonii Taylor—Mexico
U. hirta Klein ex Link—India, S.E. Asia, Australia
U. hirtella St. Hil. = *U. amethystina*
U. hispida Lam.—Honduras to N. Brazil, Trinidad.
U. hoehnei = *U. warmingii*
U. holtzei F. Muell.—Australia
U. hookeri Lehm. = *U. snaequalis*
U. humbertiana = *U. livida*
U. humboldtii R.H. Schomburgk—Venezuela, Mt Roraima
U. humilis = *U. livida*
U. humilis Vahl. = *U. bifida*
U. huntii Taylor—Brazil
U. hydrocarpa Vahl—Cuba to Brazil
U. hydrocotyloides = *U. pubescens*
U. ibarensis = *U. livida*
U. imerinensis = *U. reflexa* var. *reflexa*
U. inequalis A. DC.—S.W. Australia
U. incerta = *U. australis*
U. incisa (A. Rich.) Alain—Cuba
U. inflata Walt.—E. N. America
U. inflexa Forsk.—Trop. Africa, Madagascar, India
U. integra Le Conte = *U. biflora*
U. intermedia Hayne—Europe, Asia, N. America
U. intermedia f. *ochroleuca* (R. Hartm)—Europe, N.W. America
U. involvens Ridl.—Malaysia, Burma, Thailand, N. Australia

U. jamesoniana Oliver—Northern S. America, Antilles, Costa Rica
U. jankae = *U. australis*
U. janthina = *U. reinformis*
U. japonica = *U. australis*
U. jaquatibensis = *U. longifolia*
U. juncea Vahl—E. USA, West Indies, Northern South America
U. kaieteurensis = *U. amethystina*
U. kalmaloensis = *U. reflexa* var. *reflexa*
U. kamienskii Muell.—Australia
U. kenneallyii Taylor—N.W. Australia
U. kerrii = *U. caerulea*
U. kimberleyensis Gardner—Australia
U. kirkii = *U. arenaria*
U. kuhlmanni = *U. trichophylla*
U. kumaonensis Oliver—Himalaya
U. laciniata St. Hil. et Gir.—Brazil
U. lagoensis = *U. breviscapa*
U. lasiocaulis F. Muell.—Australia
U. lateriflora R. Br.—Australia
U. lawsoni = *U. biloba*
U. laxa St. Hil. et Gir.—Argentina, Paraguay, Brazil
U. lazulina Taylor—India
U. lehmanni = *U. capensis*
U. leptoplectra F. Muell.—Australia
U. leptorhyncha O. Schwarz—Australia
U. letestui Taylor—Africa
U. lilacina Griff. = Prob. *U. uliginosa*
U. lilliput = *U. minutissima*
U. limosa R. Br.—Australia
U. linarioides = *U. welwitschii*
U. lindmanii = *U. amethystina*
U. lingulata = *U. prehensilis*
U. livida E. Mey.—Ethiopa to Cape Province, Madagascar, Mexico
U. lloydii Merl—S. and C. America
U. lobata = *U. livida*
U. longecalcarata = *U. livida*
U. longeciliata A. DC.—Suriname, Guyana, Venezuela, Colombia, N. Brazil
U. longifolia Gardn.—Brazil
U. lundii = *U. praelonga*
U. macerrima = *U. hispida*
U. marcocheilos Taylor—Trop. Africa
U. macrorhiza Le Conte—temp. N. America, temp. E. Asia
U. macrorhyncha = *U. biflora*
U. macrophylla Masamune et Syozi = *U. uliginosa*
U. madagascariensis = *U. livida*
U. magnavesica = *U. reflexa* var. *reflexa*
U. magnifica = *U. myriocista*
U. maguirei = *U. calycifida*
U. mairii = Prob. *U. australis*
U. major = *U. australis*

U. malmeana = *U. cucullata*
U. mannii Oliver—Bamenda Highlands, Cameron Mts., Gulf of Guinea
U. marcelliana = *U. subulata*
U. mauroyae = *U. livida*
U. maxima = *U. reniformis*
U. menziesii R. Br.—W. Australia
U. meyeri Pilger—Brazil
U. micrantha = *U. erectiflora*
U. microcalyx Taylor—Trop. Africa
U. microcarpa = *U. pusilla*
U. micropetala SM.—W. trop. Africa
U. minima = *U. olivacea*
U. minor L.—Europe, Asia, N. America
U. minor f. *natans* Komiya—Japan
U. minor f. *stricta* Komiya—Japan
U. minutissima Vahl.—India to Borneo, Australia, Japan, New Guinea
U. mirabilis Taylor—Venezuela
U. mixta = *U. foliosa*
U. modesta = *U. amethystina*
U. monanthos Hook. F.—Australia, Tasmania, New Zealand
U. moniliformis Taylor—Sri Lanka
U. monophylla = *U. arenaria*
U. montana = *U. alpina*
U. moorei F.E. Lloyd = *U. dichotoma*
U. muelleri Kam.—N. Australia, New Guinea
U. multicaulis Oliver—China, Himalaya
U. multifida R. Br.—S.W. Australia
U. multispinosa = *U. minor*
U. muscosa = *U. calycifida*
U. myriocista St. Hil. et Gir.—S. America
U. nagurai = *U. exoleta*
U. nana St. Hil. et Gir.—Brazil, Guyana, Suriname, Venezuela
U. naviculata P. Taylor—Brazil, Venezuela
U. neglecta = *U. australis*
U. nelumbifolia Gardn.—Brazil
U. neottioides St. Hil et Gir.—Brazil, Colombia, Venezuela, Bolivia
U. nepalensis = *U. minor*
U. nephrophylla Benj.—Brazil
U. nervosa G. Webber ex Benj—S. America
U. nigrescens Sylven—Brazil
U. nigricaulis = *U. minutissima*
U. nipponica Makino = *U. minutissima*
U. nipponica f. *albiflora* Komiya = *U. minutissima*
U. nivea = *U. caerulea*
U. novae-zelandiae Hook. F.—New Zealand, New Cal.
U. obsoleta = *U. subulata*
U. obtusa = *U. gibba*
U. obtusiloba = *U. caerulea*
U. occidentalis = *U. ochroleuca*

U. ochroleuca R. Hartm.—Eu, N.W. America
U. odontosepala Stapf—Malawi, Zambia, Zaire
U. odontosperma = *U. livida*
U. odorata Pellegr.—S.E. Asia
U. ogmosperma = *U. reticulata*
U. oligocista = *U. subulata*
U. oligosperma = *U. foliosa*
U. olivacea Wright ex Griseb.—E. USA, S. America
U. oliverana Steyerm.—Venezuela, Amazon, Colombia
U. oliveri = *U. inflexa*
U. ophirensis = *U. caerulea*
U. orbiculata = *U. striatula*
U. orinocensis = *U. simulans*
U. ostenii = *U. tridentata*
U. pachyceras = *U. singeriana*
U. pallens = *U. obtusa*
U. panamensis Steyerm. ex Taylor—Panama
U. papillosa = *U. pubescens*
U. paradoxa = *U. spiralis* var. *spiralis*
U. parkeri = *U. capensis*
U. parkeriana = *U. obtusa*
U. parthenopipes Taylor—Brazil
U. pauciflora = *U. exoleta*
U. pauciflora = *U. caerulea*
U. peckii = *U. guyanensis*
U. pectinata = *U. longeciliata*
U. peltata = *U. pubescens*
U. peltatifolia = *U. pubescens*
U. pentadactyla Taylor—Ethiopia to Malawi and Zimbabwe
U. peranomala Taylor—China
U. perpusilla = *U. subulata*
U. pervesa Taylor—Mexico
U. philetas = *U. striatula*
U. petersoniae Taylor—Mexico
U. picta = *U. praelonga*
U. physoceras Taylor—Brazil
U. pierrei Pellegr.—S.E. Asia
U. pilifera = *U. reflexa* var. *reflexa*
U. platensis Speg.—Argentina, Uruguay
U. platyptera = *U. reflexa* var. *reflexa*
U. pobeguinii Pellegrin—Africa
U. pobeguinii = *U. spiralis* var. *pobeguinii*
U. poconensis Fronn Triata—Argentina, Brazil
U. podadena Taylor—S. Nyasaland, N.W. Mozambique
U. polygaloides Edgew—West Bengal, India
U. polyschista = *U. praelonga*
U. porphyrophylla = *U. incisa*
U. praelonga St. Hil.—Brazil
U. praeterita Taylor—India
U. praetermissa Taylor—C. America
U. prehensilis E. Mey—Ethiopia to Zimbabwe and Angola, S. Africa, Madagascar
U. protrusa Hook. F. = *U. australis*

U. pterocalycina O. Schwarz = *U. odorata* or *U. involvens*
U. pterosperma Edgew. = *U. exoleta*
U. puberula = *U. pubescens*
U. pubescens SM.—Northern S. Africa, trop. Africa, India
U. pulcherrima Sylven = *U. myriocista*
U. pulchra Taylor—New Guinea
U. pumila Benj. non Walt. = *U. subulata*
U. pumila Walt = Prob. *U. biflora*
U. punctata Wall. ex A. DC.—India, Burma, Thailand, Borneo
U. punctifolia = *U. amethystina*
U. purpurea Walt.—N. America, Cuba, Honduras
U. purpurea f. *alba* Hellquist—USA
U. purpureo caerulea St. Hil.—Brazil
U. pusilla Vahl.—C. and S. America, West Indies
U. pygmaea R. Br. = *U. minutissima*
U. quadricarinata = *U. prehensilis*
U. quelchii N.E. Brown—Venezuela, Guyana
U. quinquedentata F. Muell. ex Taylor—N. Australia
U. racemosa Wall = *U. caerulea*
U. racemosa f. *leucantha* = *U. caerulea*
U. radiata Small—N. America
U. ramosa = *U. bifida*
U. raynalii Taylor—Sudan, Senegal, Barkina Faso
U. reclinata = *U. aurea*
U. recta Taylor—Himalayas
U. recurva = *U. bifida*
U. reflexa Oliver—Senegal to S.W. and S. Africa, Madagascar
U. regnellii = *U. pubescens*
U. rehmannii = *U. capensis*
U. rendlei = *U. subulata*
U. reniformis St. Hil.—Brazil
U. resupinata Greene—E. Canada and E. USA to Northern S. America
U. reticulata S.M.—India, Ceylon
U. rhododactylos Taylor—N. Australia
U. riccioides = *U. exoleta*
U. rigida Benj.—W. trop. Africa
U. robbinsii = *U. fibrosa* E. Descr.
U. rogersiana = *U. punctata*
U. roraimensis = *U. amethystina*
U. rosea = *U. caerulea*
U. rosulata = *U. striatula*
U. rotundifolia = *U. tricolor*
U. rubra = *U. tridentata*
U. rubricaulis = *U. guyanensis*
U. saccata Le Conte = *U. purpurea*
U. saccata Merl ex Luetzelb. = *U. subulata*
U. sacciformis Benj. = *U. australis*
U. salwinensis Hand.-Mazz.—China
U. salzmanni = *U. hydrocarpa*
U. sandersonii Oliver—S. Africa
U. sandwithii Taylor—Guyana, Suriname, Venezuela

U. sanguinea = *U. livida*
U. saudadensis = *U. geminoloba*
U. scandens Benj.—trop. Africa, Madagascar, trop. Asia, Australia
U. scandens var. *firmula* Oliv.
U. schimperi = *U. jamesoniana*
U. schinzii = *U. capensis*
U. schultesii Fernandez Perez—Colombia
U. schweinfurthii = *S. scandens* ssp. *schweinfurthii*
U. sciaphila = *U. pubescens*
U. sclerocarpa = *U. juncea*
U. secunda = *U. obtusa*
U. sematophora = *U. livida*
U. siakujiiensis = *U. australis*
U. siamensis = *U. minutissima*
U. simplex R. Br.—Australia
U. simplex = *U. juncea*
U. simulans Pilger—Florida to Brazil, trop. Africa
U. singeriana F. Muell.—Australia
U. sinuata = *U. amethystina*
U. smithiana Wight—India
U. sootepenis = *U. caerulea*
U. spartea = *U. livia*
U. spatulifolia = *U. amethystina*
U. spicata = *U. erectiflora*
U. spiralis SM.—Trop. Africa
U. spirandra = *U. obtusa*
U. sprengelii = *U. capensis*
U. spruceana Benth ex Oliv—S. America
U. squamosa = *U. graminifolia*
U. stanfieldii Taylor—W. Africa
U. steenisii Taylor—Sumatra
U. stellaris L.F.—S. Africa, Madagascar, trop. Asia, Australia
U. steyermarkii Taylor—Venezuela
U. striata Le Conte ex Torrey—N. America
U. striatula SM.—Trop. Africa, India to New Guinea
U. stricta = *U. juncea*
U. stricticaulis (Koenig) Stapf ex Gamble = *U. polygabides*
U. subpeltata = *U. pubescens*
U. subrecta = *U. graminifolia*
U. subsimilis = *U. novae-zelandiae*
U. subulata L.—Nova Scotia to Argentina, trop. Africa, Madagascar, Thailand, Borneo, Portugal
U. sumatrana = *U. exoleta*
U. taikankoensis = *U. striatula*
U. tenella R. Br.—S. and SW Australia
U. tenerrima = *U. baouleensis*
U. tenuicaulis = *U. vulgaris* f. *tenuicaulis*
U. tenuifolia = *U. obtusa*
U. tenuis = *U. obtusa*
U. tenuiscapa = *U. subulata*
U. tenuissima Tutin—Northern S. America
U. tepuiana = *U. amethystina*
U. ternata = *U. tridentata*

U. terrae-reginae Taylor—Australia
U. tetraloba Taylor—W. Africa
U. *thomasii* = *U. pubescens*
U. *tinguensis* = *U. obtusa*
U. tortilis Welw. ex Oliver—Africa
U. *transrugosa* = *U. livida*
U. *tribacteata* = *U. arenaria*
U. trichophylla Spruce ex Oliver—Brazil, Venezuela, Guyana
U. *trichoschiza* = *U. stellaris*
U. tricolor St. Hil.—Brazil, Paraguay, Colombia, Venezuela
U. *tricrenata* = *U. obtusa*
U. tridactyla Taylor—W. Australia
U. tridentata Sylven—S. Brazil, Uraguay, Argentina
U. triflora Taylor—N. Australia
U. triloba Benj.—S. America
U. *triloba* Good = *U. subulata*
U. *trinervia* = *U. amethystina*
U. *triphylla* = *U. geminiloba*
U. troupinii Taylor—Trop. Africa
U. tubulata F. Muell.—Australia
U. *turumiquirensis* = *U. amethystina*
U. uliginosa Vahl.—India to Australia
U. uniflora R. Br—Australia
U. unifolia Ruiz et Pavon—S. and C. America
U. *vaga* = *U. hydrocarpa*
U. *velascoensis* = *U. amethystina*
U. *venezuelana* = *U. pubescens*

U. *verapazensis* = *U. jamesoniana*
U. *verticillata* Benj. = *U. biloba*
U. *villosula* = *U. benjaminiana*
U. violacea R. Br.—Australia
U. *virgatula* = *U. juncea*
U. viscosa Spruce et Oliver—Northern South America
U. vitellina Ridl.—Malaysia
U. volubilis R. Br.—Australia
U. *vulcanica* = *U. australis*
U. vulgaris L.—N. temp. region incl. Europe, N. Africa, Temp. Asia
U. *vulgaris* f. *fixa* Komiya—Japan
U. *vulgaris* f. *tenuicaulis* Miki et Komiya—Japan
U. *wallichiana* = *U. scandens* var. *firmula*
U. warburgii Goebel—China
U. warmingii Kam.—S. America
U. welwitschii Oliver—Katanga, Rwanda, Burundi to Angola, S. Africa, Madagascar
U. westonii Taylor—W. Australia
U. wightiana Taylor—India
U. *williamsii* = *U. amethystina*
U. *yakusimensis* = *U. uliginosa*

This plant list builds upon information provided in 'A World Carnivorous Plant List,' *Carnivorous Plant Society Newsletter*, Vol 15, nos 3 & 4, December 1986.

KEY TO WORLD PLANT LIST

This list includes all carnivorous plant species known at the time of publication. As this is such a dynamic area, it is likely that the list will expand, or even decrease in some areas as the 'carnivorous' nature of some species are more fully investigated. A number of conventions have been followed, to make this list more wieldy, and they are as follows:

- Both *genus* and *species* names are in italics, following scientific practice
- x Indicates that the species is a hybrid
- ssp Indicates that the plant is a subspecies. Subspecies often refer to different geographical locations for the same species, or some significantly variant characteristic
- f Indicates a variation of form within a subspecies
- var. Indicates that there is a variation in a species at the same location; varieties may have different forms

- = Indicates a synonym, and the final name is that currently accepted
- ( ) Directly after the species name contains the name of the person who first published or officially 'discovered' it
- ex When placed between two names it indicates that the first person discovered or created (in the case of hybrids) the species, while the second person published it
- (L) After the name of the species indicates that Carl Linnaeus named it, and published it in his *Species Plantarum*
- Origin Following the name is the place of origin of the species (where a hybrid occurs naturally, the place of origin is also noted)
- Hort. = Horticultural hybrid

References

The following titles are useful for taxonomical information: for more detailed references on carnivorous plants, consult the Bibliography.

Allan, H.H. *Flora of New Zealand*, Wellington, New Zealand, Government Printer, 1961.

Bailey, F.M. *The Queensland Flora*, Brisbane, H.J. Diddams and Co., 1900.

Bailey, L.H. *The Standard Cyclopedia of Horticulture*, 1919.

Bentham, G. *Flora Australiensis: A Description of the plants of the Australian Territory*, London, L. Reeve and Co., 1864.

Biota of N. America, University of North Carolina Press, 1980.

Black, J.M. *Flora of South Australia*, Adelaide, South Australia, Government Printer, 1965.

Carnivorous Plant Newsletter, 1974 & 1986.

Engler, A. *Das Pflanzenreich*, 1908.

Erickson, R. *Plants of Prey in Australia*, Lamb Publications, 1968.

Flora of Tropical East Africa, Kew, Royal Botanic Gardens, 1973.

Flore d'Afrique Centrale: Zaire—Rwanda—Burundi, Belgium, Bruxelles, Jardin Botanique Nationale, 1972.

Gleason, H.A. & Conquist, A. *Manual of the Vascular Plants of the N.E. States and Adjacent Canada*, New York, Van Nostrand, 1963.

Juniper, B.E., Robins, R.J. & M. Joel. *The Carnivorous Plant*, London, Academic Press, 1989.

Laundon, J.R. *Flora of Tropical East Africa*, 1959.

Lentibulariaceae, New York Botanical Gardens, 1967.

Lentibulariaceae: Flore d'Afrique Central, Zaire—Rwanda—Burundi.

Makino, T. *An Illustrated Flora of Nippon with the Cultivated and Natural Plants*, Hokuryukwen, 1860–62.

Marchant, N.G., Aston, H.L. & George, A.S., *Flora of Australia*, Australian Government Publications, 1982.

Mueller, F., *The Plants Indigenous to the Colony of Victoria*, Melbourne, Bailliere.

Small, J.K. *Flora of the Southeastern United States*, New York, 1913.

Spermatophytes, Jardin Botanique National de Belgique, Bruxelles, 1972.

Steyermark, J. *Fieldina: Botany*, 1952.

Taylor, Peter *The Genus* Utricularia—a *Taxonomic Monograph*, Kew Bulletin, Royal Botanic Gardens, Kew, 1989.

Recommended Reading

For other works, please consult the bibliography overleaf.

Cheers, G. *Carnivorous Plants*, Carnivor and Insectivor Plants, Victoria, Australia, 1983.

Cheers, Gordon & Julie Silk. *Carnivorous Plants* (a children's guide), Puffin Books, Ringwood, Victoria, 1992.

Darwin, Charles. *Insectivorous Plants*, London, John Murray, 1876.

Kondo, K. & M. *Carnivorous Plants of the World in Colour*, Tokyo, Lenohikari Association, 1983.

Lloyd F.E. *The Carnivorous Plant*, New York, Dover Publications Inc., 1976.

Lowrie, A. *Carnivorous Plants of Australia*, Vol. 1–111, Perth, University of Western Australia Press, 1987, 1989, 1992.

Nepenthes of *Mount Kinabalu*, Sabah National Park.

Pietropaolo, J. & P. *Carnivorous Plants of the World*, Oregon, Timber Press, 1986.

Schnell, D. *Carnivorous Plants of the United States and Canada*, North Carolina, John F. Blair, 1976.

Organisations

Carnivorous plant societies exist all over the world, and vary in size and approach: these societies can prove valuable sources of information for both amateurs and professionals, as well as providing useful networks for obtaining plants and seeds, depending on export laws. The list provided here is by no means exhaustive, but as personnel and contacts for these societies tend to change, it is a good idea to follow up contacts in your own country.

The American/International Carnivorous Plant Society publishes a newsletter full of useful contacts and articles. Botanical gardens and museums are also good sources of information, especially when trying to identify a plant.

International Carnivorous Plant Society
The Fullerton Arboretum
Department of Botany
California State University
Fullerton California 92634
United States of America

Australian Carnivorous Plant Society
PO Box 256
Goodwood South Australia 5034
Australia

Carnivorous Plant Society England
174 Baldwins Lane
Croxley Green
Hertfordshire WD3 3LQ
England

Carpass
Milan Beutelheuser
UL, SND 30
Ivank Pri Dunaji
90 2B
Czechoslovakia

Dutch Carnivorous Plant Society
Krayenthoeffstraat 51
Amsterdam 1018-RH
Holland

French Carnivorous Plant Society
41 rue Henri II Plantagenet
Rouen F-76100
France

Gesellschaft für Fleischfressende Pflan
Hohlwreg 15B
Nudlinge D 8738
Germany

Insectivorous Plant Society of Japan
c/o Nippon Dental College
Fujima, Chiyoda-ku Tokyo 102
Japan

New Zealand Carnivorous Plant Society
PO Box 162
Christchurch
New Zealand

Victorian Carnivorous Plant Society
PO Box 8
Greensborough Victoria 3088
Australia

BIBLIOGRAPHY

Adams, Koenigsberg & Langhans, 'In vitro propagation on *Cephalotus follicularis*, (Australian Pitcher Plant)', *Horticultural Scientist*, 14: 512-513, 1979.

Airy Shaw, H.K., 'On the *Dioncophyllaceae*, a remarkable new family of flowering plants', *Kew Bulletin*, 3, 1951.

Allen, Oliver E., 'Linnaeus', *Horticulture*, 30-36, January 1985.

Angerillia, N.P.D. & Beirne, B.P., 'Influences of aquatic plants on colonization or artificial ponds by mosquitoes and their predators', *Canadian Entomologist*, 112: 793-796, 1980.

Aston, Helen, *Aquatic Plants of Australia*, Melbourne, Melbourne University Press, 1973.

Bailey, F.M., *The Queensland Flora*, Vol. IV, Brisbane, Queensland Government, 1901.

Bailey, L.H., *The Standard Cyclopedia of Horticulture* IV, Vol. 1, 2 & 3, 1919.

Bailey, L.H., *The Nursery Manual*, 1947.

Baines, T., *Greenhouse and Stove Plants*, London, John Murray, 1885 & 1894.

Beebe, J.D., 'Morphogenetic responses of seedlings and adventitious buds of the carnivorous plant *Dionaea muscipula* in aseptic culture', *Botanical Gazette*, 141: 396-400, 1980.

Bentley, Robert, *A Manual of Botany*, London, J & A Churchill, 1873.

Berenbaum, May, 'Pests—Scale', *Horticulture*, 20-22, January 1985.

Bingham, Madeleine, *The Making of Kew*, London, Michael Joseph.

Bower, F.O., 'On the pitcher of *Nepenthes*: A study in the morphology of the leaf', *Annals of Botany* V. III, 1889-90.

British Association for the advancement of Science, *Handbook & Guide to W.A.*, Perth, 1914.

Britten, W.J., 'Sir Harry James Veitch', *Kew Bulletin*, 300-301, 1924.

Burbidge, F.W., *The Gardens of the Sun*, London, John Murray, 1880.

Carnivorous Plant Newsletter, California, The Fullerton Arboretum, 1973-1984.

Carron, W., *Narrative of an Expedition Undertaken Under the Direction of Mr Assistant Surveyor E.B. Kennedy etc.*, Sydney, Femp & Fairfax, 1849.

Casper, S. Jost, 'On *Pinguicula macroceras* link in North America, *Rhodora* 64/65: 212-221, 1962/63.

Ceska, A. & Bell, M.A.M., '*Utricularia* in the Pacific Northwest', *Madrono* 21/22, 213-220, 1971/74.

Chapman, A.D., 'The properties of *Drosera binata*', *Australian Plants*, 11: 175-182, 1981.

Conn, B.J., '*Drosera peltata—Drosera auriculata* complex', *Adelaide Botanical Garden*, 3(1), 91-100, 1981.

Corner, E.J.H., 'Mt Kinabalu history', *Proceedings of the Royal Society* (London), Series B, No. 161, 1964-65.

Danser, B.H., 'The *Nepenthaceae* of the Netherlands Indies', *Bulletin de Buitenzorg*, Series III, Vol. IX, 249-438, 1929.

Darwin, Charles, *Insectivorous Plants*, London, John Murray, 1876.

Darwin, Francis, *The Life and Letters of Charles Darwin*, Vol. 1, 2 & 3, London, John Murray, 1888.

Darwin, Francis, & Seward, A.C., *More Letters of Charles Darwin*, Vol. 1 & 2, London, John Murray, 1903.

De Beer, Sir Gavin, *Charles Darwin, Evolution by Natural Selection*, Thomas Nelson & Sons, Ltd, 1963.

De Buhr, L.E., 'Two new species of *Drosera* from W.A.', *Aliso*, 8(3): 263-271, 1975.

Dixon & Pate, 'Phenology, morphology and reproductive biology of the tuberous Sundew *Drosera erythrorhiza*', *Australian Journal of Botany*, 26: 441-464, 1978.

Dixon, K.W., '*Drosera*', *Australian Plants*, 11: 170-175, June 1981.

Dreir, David, 'Venus's Flytrap case closed', *Omni*, July 1983.

Duddington, C.L., 'Fungi that attack microscopic animals', *The Botanical Review*, XXI No. 7, 377-439, July 1955.

Erickson, R., 'Australian carnivorous plants', *Australian Plants*, 3: 319-322, June 1966.

Erickson, R., 'Bladderworts—Fairy Aprons', *Australian Plants*, 4: 292-293, June 1968.

Erickson, R., 'The plants that prey on living creatures', *Australian Plants*, 4: 287-291.

Erickson, R., *Plants of Prey in Australia*, Perth, Lamb Paterson, 1968.

Erickson, R., *Triggerplants*, Perth, University of Western Australia Press, 1981.

Fairburn, D.C. & Pring, G.H., '*Nepenthes* (Pitcher Plants)', *Missouri Botanic Garden Bulletin*, 31: 160-185, 1943.

Finch, J., 'Tailored genes', *Horticulture*, 50-55, April, 1984.

Fish, D., 'Insect-plant relationships of the insectivorous pitcher plant *Sarracenia minor*', *Florida Entomologist*, Vol. 59, No. 2, 199-203.

Fish, D. & Beaver, R.A. 'A bibliography of the aquatic fauna inhabiting Bromeliads (*Bromeliaceae*) and Pitcher Plants (*Nepenthaceae* and *Sarraceniaceae*)', *Proceedings of the Florida Anti-Mosquito Association*, 49th Meeting, April 2-5, 1978.

Fish, D., & Hall, D.W., 'Succession and stratification of aquatic insects inhabiting the leaves of the insectivorous Pitcher Plant', *Sarracenia purpurea*, *The American Midland Naturalist*, Vol. 99, No. 1, 172-183, January, 1978.

Folkerts, G.W., 'The gulf coasts Pitcher Plant bogs', *American Scientist*, 70: 260-267, May-June, 1982.

Forsyth, A., 'Bog behavior Pitcher Plants and Sundews', *Horticulture*, 60: 24–29, October 1982.

Franck, D.H., 'The morphological intepretation of epiascidiate leaves', *Botanical Review*, Vol. 42 No. 3, 345–388, July–September, 1976.

Gardener, G.A., *Wildflowers of West Australia*, 11th ed., Perth, WA Newspapers Ltd., 1973.

Gardiner, W., 'On the power of contractility exhibited by the protoplasm of certain plant cells', *Annals of Botany*, 362–367, 1887–89.

Gibson, R.W. & Turner, R.H., 'Insect-trapping hairs on potato plants', *Pest Articles and News Summaries*, 23 (3): 272–277, 1977.

Green, Sally, 'Notes on the distribution of *Nepenthes* species in Singapore', *Gardens' Bulletin*, Singapore, XXII, 53–65, 1967.

Green, Sally, Green, T.L., & Heslop Harrison, 'Seasonal heterophylly and leaf gland features in *Triphyophyllum* (*Dioncophyllaceae*), a new carnivorous plant genus', *Botanical Journal of the Linnean Society*, 78: 99–116, February 1979.

Gluck, Prof. D., 'Contributions to our knowledge of the species of *Utricularia* of Great Britain with special regard to the morphology and geographical distribution of *Utricularia ochroleuca*', *Annals of Botany*, Vol. XXVII, 607–620, 1913.

Grieve, Prof. B.J., 'The Pitcher Plants as an insect trap', *Australian Plants*, 4: 28, December 1966.

Hamilton, A.G., 'Notes on the West Australian Pitcher Plant', *Proceedings NSW Linnean Society*, 36–52, 1904.

Hartmann, H.T. & Kester, D.E., *Plant Propagation, Principles and Practices*, 3rd ed., New Jersey, Prentice-Hall Inc., 1959.

Harvey, Julian & Kettlewell, H.B.D., *Charles Darwin and His World*, Thames & Hudson, 1965.

Henderson, D.M. & Prentice, H.T., *International Directory of Botanical Gardens*, 3rd Ed., Utrecht-Netherlands, 1977.

Heslop-Harrison, Y., 'Carnivorous plants a century after Darwin', *Endeavor*, 35/126: 114–122, 1976.

Heslop-Harrison, Y. & Knox, R.B. 'A cytochemical study of the leaf-gland enzymes of insectivorous plants of the genus *Pinguicula*', *Planta*, 96: 183–211, 1971.

Hibberd, Shirley, *The Amateur's Greenhouse and Conservatory*, London, Groombridge & Sons, 1875.

Insectivorous Plant Society of Japan, *Carnivorous Plants*, Japan.

Insectivorous Plant Society of Japan, *The Journal of Insectivorous Plant Society*, 1983–84.

Jabs, C., 'Bringing in the seeds', *Horticulture*, 35–38, June 1984.

Jacobson, S.K., *A Guide to Kinabalu National Park*, Sabah National Parks Publication No. 3, 1979.

Jacobson, S.L., & Benolken, R.M., 'Response properties of a sensory hair excised from Venus's Flytrap', *Journal of General Physiology*, 56: 64–82.

Juniper, B.E., Robins, R.J., & Joel, D., *The Carnivorous Plant*, London, Academic Press, 1989.

Kisha, J.S., 'Observations on the trapping of the whitefly *Bemisia tabaci* by glandular hairs on tomato leaves', *Annals of Applied Biology*, 97: 123–127, 1981.

Kondo, K. & M., *Carnivorous Plants of the World in Colour*, Tokyo, Lenohikari Association.

Kondo, K., 'Comparison of *Utricularia cornuta* and *Utricularia juncea*', *American Journal of Botany*, 59: 23–37, 1972.

Lecoufle, Marcel, *Carnivorous Plants: Their Care and Cultivation*, Blandford Press, 1990.

Lichtner, F.T., & Williams, S.E., 'Prey capture and factors controlling trap narrowing in *Dionaea* (Droseraceae)', *American Journal of Botany*, V. 64(7): 881–886, 1977.

Lloyd, Francis E., '*Drosera* numbs collector's Fingers', *Victorian Naturalist*, 64: 35, July 1937.

Lloyd, Francis E., 'Is *Roridula* a carnivorous plant?', *Canadian Journal of Research*, 10: 780–86, 1934.

Lloyd, Francis E., 'Notes on *Utricularia*, with special reference to Australia, with descriptions of four new species', *Victorian Naturalist*, 64: 91–166, October 1936.

Lloyd, Francis E., *The Carnivorous Plants*, New York, Dover Publications, Inc., 1976.

Loder, G., 'Life and Letters of Sir J.D. Hooker', *Kew Bulletin*, 32: 345–351, 1918.

Lowrie, A., *Carnivorous Plants of Australia*, Vol. I–III, Perth, University of Western Australia Press, 1987, 1989, 1992.

Lullfitz, F., 'The West Australian Pitcher Plant (*Cephalotus follicularis*)', *Australian Plants*, 4: 34–37, December 1966.

McKinney, K.B., 'Physical characteristics on the foliage of beans and tomatoes that tend to control some small insect pests', *Journal of Economic Entomology*, Vol. 31 No. 5, 630–631, 1938.

Marchant, N.G., Aston, H.I., & George, A.S., *Droseraceae, Flora of Australia*, Vol. 8, Australian Government Publication, 1982.

Meeuse, B.J.D., 'The Voodoo Lily', *Scientific American*, Vol. 215, 80–88, 1966.

Metcalfe, C.R., 'The anatomical structure of the *Dioncophyllaceae* in relation to the taxonomic affinities of the family', *Kew Bulletin*, 1951.

Moser, D., 'Big thicket of Texas', *National Geographic*, 504–528, October 1974.

Newman, G.H., 'Cultivation and propagation of insectivorous plants', *New England Wildflower Notes*, Spring 1974.

Nigel Hepper, F. (ed.), *Kew Gardens for Science and Pleasure*, London, Her Majesty's Stationery Office, 1982.

Osborn, C.S., *Madagascar, Land of the Man-Eating Tree*, New York, Republic Publishing Co., 1924.

Overbeck, C., *Carnivorous Plants*, Minneapolis, Lerner Publications Co., 1982.

Pietropaolo, J. & P., *The World of Carnivorous Plants*, New York, R.J. Stoneridge, 1974.

Pietropaolo, J. & P., *Carnivorous Plants of the World*, Oregon, Timber Press, 1986.

Plummer, G.L., 'Soils of the Pitcher Plant habitat', *Ecology*, v. 44, 727–734, 1963.

Plummer, G.L. & Kethley, J.B., 'Foliar absorption of amino acids, peptides, and other nutrients by the Pitcher Plant, *Sarracenia flava*', *Botanical Gazette*, 125(4), 245–260, 1964.

Reinert, G.W. & Godfrey, R.K., '*Utricularia inflata* and *Utricularia radiata*', *American Journal of Botany*, Vol. 49: 213–220, 1962.

Richards, P.W., *Tropical Rain Forest*, London, Cambridge University Press, 1964.

Ridley, H.N., 'On the foliar organs of a new species of *Utricularia* from St Thomas, West Africa', *Annals of Botany*, Vol. II, 305–307, 1888–89.

Robins, R.J., 'The nature of the stimuli causing digestive juice secretion in *Dionaea muscipula* Ellis (Venus's Flytrap)', *Planta*, 128, 263–265, 1976.

Robins, R.J. & Juniper, B.E., 'The secretory cycle of *Dionaea muscipula* Ellis, II. Storage and synthesis of the secretory proteins', *New Phytologist*, Vol. 86 no. 3: 297–311, 1980.

Russell, M.C., 'Growing tuberous Sundews', *Australian Plants*, 14: 290–291, June 1968.

Sabah Society, *Kinabalu, Summit of Borneo*, Malaysia, The Sabah Society, 1978.

Salisbury, E.J., 'On the occurrence of vegetative propagation in *Drosera*', *Annals of Botany*, Vol. XXIX, 308–110, 1915.

Scala, J., Iott, K., Schwab, D.W., Semersky, F.E., 'Digestive secretion of *Dionaea muscipula*', *Plant Physiologist*, 44: 367–371, 1969.

Schnell, D., *Carnivorous Plants of the United States and Canada*, North Carolina, John F. Blair, 1976.

Schnell, D., 'More about the sunshine pitchers', *NYBG Garden Journal*, 14(5): 146–147, 1974.

SGAP—Queensland, *A Horticultural Guide to Austrtalian Plants* Set 7, The Society for Growing Australian Plants, August 1980.

Shivas, Roger, *Pitcher Plants of Peninsular Malaysia and Singapore*, Maruzen, 1984.

Simons, P., 'The touchy life of nervous plants', *New Scientist*, March 1982.

Simons, P.J., 'The role of electricity in plant movements', *New Phylologist*, Vol. 87, 11–37, 1981.

Skutch, A.F., 'The secret of the bladderwort', *Scientific American*, December, 1928.

Smee, Alfred, *My Garden*, 1872.

Smythies, B.E., *The Distribution and Ecology of Pitcher Plants (Nepenthes) in Sarawak*, Conservatory of Forestry.

Sprague, T.A., *Dioncophyllum, Kew Bulletin*, No. 4, 1916.

Stanley, T.D., '*Nepenthaceae*', *Flora of Australia*, Vol. 8, 7–9, Australian Government Publication, 1982.

Steyermark & Smith, 'A new *Drosera* from Venezuela', *Rhodora*, 76: 491–493.

Taylor, P., 'The genus *Utricularia* in Africa and Madagascar', *Kew Bulletin*, 18(1): 1–248, 1964.

Taylor, P., 'The genus *Utricularia*: a taxonomic monograph,' *Kew Bulletin*, Royal Botanic Gardens, Kew, 1989.

Taylor, S., 'Insectivorous plants in British Columbia', *Davidsonia*, 10: 41–53, 1979.

Turnbull, J.R. & Middleton, A.T., *A Preliminary Review of the Sabah Species of Nepenthes*, Canada, Department of Botany & Genetics, University of Guelph, 1981.

Turnbull, J.R. & Middleton, A.T., 'Three new *Nepenthes* from Sulawesi Tengah', *Reinwardtia*, Vol. 10, Pt 2, 107–111, 1984.

Turrill, W.B., *The Royal Botanic Gardens, Kew, Past and Present*, London, Herbert Jenkins Ltd., 1959.

Veitch, James H., '*Nepenthes*', *Journal of the Royal Horticultural Society*, Vol. XXI, 226–262, 1897–1898.

Veitch, James H., *Hortus Veitchii—A History*, London, James Veitch & Sons Ltd., 1906.

Vines, S.H., 'On the digestive ferment of *Nepenthes*', *Biological Journal of the Linnean Society*, 15: 427–431, 1876–77.

Vines, S.H., 'The proteolytic enzyme of *Nepenthes*', *Annals of Botany*, 1897, 1898, 1901.

Wheeler, G.A. & Glaser, P.H., 'Vascula plants of the Red Lake Peatland, Northern Minnesota', *Michigan Botanist*, 21: 89–93, 1982.

White, C.T., 'A new type of Sundew from North Queensland', *Victorian Naturalist*, 57: 94–95, September 1940.

White, J.W., 'In the greenhouse—pest control without pesticides', *Horticulture*, 52–54, November, 1984.

Whitehead, B., 'Fairy Fans', *Australian Plants*, 6: 344–347, September 1972.

Williams, S.E., 'Comparative sensory physiology of the Droseraceae—the evolution of a plant sensory system', *American Philosophical Society Proceedings*, v. 120, No. 3, 187–204, 1976.

Williams, S.E., & Bennett, A.B., 'Leaf Closure in the Venus Flytrap: An Acid Growth Response', *Science*, Vol. 218, 1120–1122, 1982.

Willims, S.E., & Pickard, B.G., 'The Role of Action Potentials in the Control of Capture Movements of *Drosera* and *Dionaea*', *Plant Growth Substances*, 470–480, 1979.

Williams, S.E., & Pickard, B.G., 'Connections and Barriers between Cells of *Drosera* Tentacles in Relation to Their Electrophysiology', *Planta*, 1974.

Willis, J.H., '*Cephalotus follicularis*', *Western Australian Naturalist*, Vol. 10 No. 1, November 2 1965.

Willis, J.H., Fuhner, B.A., & Rotherham, E.R., *Field Guide to the Flora and Plants of Victoria*, 1975.

Windler, D.R., 'A water plant with a taste for tiny beasts', *Smithsonian*, 10: 91–94, May 1979.

Winston & Gorham, 'Turions and dormancy states in *Utricularia vulgaris*', *Canadian Journal of Botany*, 57: 2740–2749, 1979.

Woodcock, K., *Carnivorous Plants*, Cambridge University Botanic Garden, 1979.

Yanchinski, S., 'De Fungus', *New Scientist*, 38, October 1980.

Zahl, Paul A., 'Plants that eat insects', *National Geographic*, 119: 643–659, 1961.

Zahl, Paul A., 'Malaysia's giant flowers and insect-trapping plants, *National Geographic*, May 1964.

Photographic Credits

Illustrations
All black and white illustrations © Margaret Hodgson.

Photographs
Photographs are in page order: l (left), r (right), t (top), b (bottom), m (middle). Species names have been used to identify photographs on pages 18-20. The full name of each photographer is used only in the first instance. All copyright remains with individual photographers or institutions.

Front Cover: Andre Martin, with thanks to Botanic Gardens, Sydney; Back cover: Mary Nemeth, Photo/Nats; p vii Hal Horwitz, Photo/Nats; p viii Tony Holdcroft; p x Horwitz, Photo/Nats; p 1 David M. Stone; p 4 Larry Pitt, with thanks to Botanic Gardens, Melbourne; p 5 Larry Pitt, with thanks to La Trobe University; p 6 BBC London; p 7 (l) Alfred Byrd Graf; p 7 (r) Pitt, with thanks to La Trobe University; p 8 (l) Dr Richard M Adams II; p 8 (r) Dr Richard M. Adams II; p 9 Darwinian Society, London; p 10 Royal Botanic Gadens, Kew; p 11 Royal Botanic Gardens, Kew; p 12 Royal Botanic Gardens, Kew; p 13 Pitt; p 14 Jacques Haldi; p 15 Pitt, with thanks to Botanic Gardens, Melbourne; p 18 *Heliamphora heterodoxa* Pitt; p 18 *Catopsis berteroniana* Lorenz Bütschi; p 18 *Sarracenia purpurea* Pitt; p 18 *Utricularia vulgaris* Richard Davion; p 18 *Dionaea muscipula* Pitt; p 18 *Nepenthes villosa* Kjell B. Sandved; p 19 *Drosera filiformis* Richard Stomberg; p 19 *Drosera peltata* Davion; p 19 *Sarracenia oreophila* Pitt; p 19 *Nepenthes lowii* Sandved; p 19 *Nepenthes masoalensis* Marcel Lecoufle; p 20 *Cephalotus follicularis* Pitt; p 20 *Nepenthes* (male flower) Patricia Pietropaolo; p 20 *Sarracenia rubra* Horwitz, Photo/Nats; p 20 *Nepenthes rajah* Sandved; p 20 *Nepenthes* (female flower) Pitt; p 20 *Sarracenia flava* Horwitz, Photo/Nats; p 20 *Drosera paleacea* Mathew Denton; p 22 A.W. Sanborn; p 23 Pitt; p 24 Pitt; p 25 Pitt; p 26 (b, r) J. Bogner; p 26 (t, r) Beckwith; p 27 Pitt; p 28 Donald Schnell; p 29 Ian Rogers; p 30 (l, t) Pitt; p 30 (l, m) Pitt; p 30 (r) Davion; p 31 (t × 3) Rogers; p 31 (b, l) Bütschi; p 31 (b, r) Davion; p 32 (b, l) Pitt; p 32 (b, r) Pietropaolo; p 33 Stomberg; p 34 Jo-ann Ordano Photo/Nats; p 35 Davion; p 36 Isamu Kusakabe; p 37 (b, r) Horwitz, Photo/Nats; p 37 (t, l) Pietropaolo; p 39 Davion; p 40 (t × 3) Pitt; p 40 (b, l) Pitt; p 40 (m, r) Adams; p 41 Rogers; p 44 Pitt; p 45 Pitt; p 46 Rogers; p 47 Pitt; p 48 Davion; p 50 Mazrimas; p 52 Bütschi; p 54 (t, r) Pitt; p 54 (b, l) Charles L. Powell II; p 56 Bütschi; p 58 Steve Beckwith; p 59 (l, t) Rogers; p 59 (b, r) Rogers; p 59 (t, r) Pitt; p 61 Pitt; p 63 Robert E. Lyons, Photo/Nats; p 65 (b, l) Beckwith; p 65 (t, r) Rogers; p 66 (b, l) Rogers; p 66 (t, r) Pitt; p 67 (t) Don Johnston, Photo/Nats; p 67 (b, l) M. Studnicka; p 68 (t, l) Horwitz Photo/Nats; p 68 (b, r) Alain Christophe; p 69 (t, l) Martin Cheek, Kew; p 69 (t, r) Davion; p 69 (b, r) Davion; p 70 (b, l) Cheek, Kew; p 70 (t, r) David M. Stone, Photo/Nats; p 71 (b, l) Rogers; p 71 (t, r) Denise Grieg; p 72 (m, l) Walter; p 72 (t, r) Dorothy S. Long, Photo/Nats; p 73 (b, l) Photo/Nats; p 73 (t, r) Stoutamire; p 74 (m, l) Rogers; p 74 (b, l) Beckwith; p 74 (t, r) Pietropaolo; p 75 (b, l) Rogers; p 75 (t, r) Davion; p 76 (b, r) Rogers; p 76 (t, r) Cheek, Kew; p 76 (m, l) Rogers; p 77 (t, r) Davion; p 77 (b, l) Denton; p 78 (t, l) Davion; p 79 (t, r) Davion; p 79 (b, r) Beckwith; p 80 (t) Holdcroft; p 80 (b) Rogers; p 81 (t, r) Davion; p 81 (b) Davion; p 83 (t) Studnicka; p 83 (b, r) Davion; p 85 (t, l) Cheek, Kew; p 85 (m, r) H. Herkner; p 85 (b, l) Bütschi; p 87 (t, r) Larry Pitt; p 87 (b, l) Stomberg; p 87 (b, r) Stephen Tong; p 88 Bütschi; p 89 (m, l) Johannes Marabini; p 89 (t, r) Cheek, Kew; p 92 (b, l) Pitt; p 92 (t, r) Cathy Law; p 93 (b, l) Beckwith; p 93 (t, r) Kusakabe; p 94 (t, l) A. Lamb; p 94 (b, r) Justin Tong; p 95 (t, l) J. Tong; p 95 (t, r) Sandved; p 95 (b, m) Geoff Roberts; p 96 (m, l) Lamb; p 96 (t, r) Pitt; p 96 (b, r) Lamb; p 97 (t, m) J. Tong; p 97 (t, r) Lamb; p 97 (b) Lamb; p 98 (t, l) Beckwith; p 98 (b, r) Sandved; p 99 Sandved; p 100 (b, l) Peter Lavarack; p 100 (t, r) A. Lamb; p 100 (b, r) Walter; p 101 (t, r) Pitt; p 101 (b, r) J. Forlonge; p 101 (m, l) Law; p 102 Pitt; p 103 (t, r) Lamb; p 103 (b, l) Lamb; p 104 (t, l) Johannes Marabini; p 104 (t, r) Lamb; p 105 (m, l) Marabini; p 105 (t, r) Lamb; p 105 (b, l) Lamb; p 105 (b, r) Forlonge; p 106 (t, l) Sandved; p 106 (b, l) Walter; p 106 (r) Walter; p 107 Sandved; p 109 (m, l) Cheek, Kew; p 109 (b, r) Studnicka; p 110 (t, l) Gordon Blanz; p 110 (b, r) Studnicka; p 111 (m, l) Davion; p 111 (b, r) Pietropaolo; p 112 (t, r) Rogers; p 112 (b, l) Les Saucier, Photo/Nats; p 112 (m, r) Blanz; p 113 (t,r,) Davion; p 113 (b, l) Davion; p 114 (t, r) Davion; p 114 Stomberg; p 115 Pitt; p 116 (b, l) Pitt; p 116 (t, r) Beckwith; p 117 (t, l) Alan Jaeger; p 117 (b, r) Pitt; p 118 (t, l) Mary Helen Nemeth, Photo/Nats; p 118 (b, r) Rogers; p 119 (t,) Saucier, Photo/Nats; p 119 (b, l) Saucier, Photo/Nats; p 120 (t, l) Saucier, Photo/nats; p 120 (r) Saucier, Photo/Nats; p 122 Bogner; p 124 (b, l) Studnicka; p 125 (t, r) Rogers; p 125 (b, l) Beckwith; p 125 (m. r) David M. Stone, Photo/Nats; p 126 (m, l) Knights; p 126 (t, r) Adrian Walter; p 126 (b, r) Stone, Photo/Nats; p 127 (t, l) Kevin Donovan; p 127 (b, r) Holdcroft; p 128 (t, r) Stoutamire; p 128 (b, l) Studnicka; p 129 (t, l) Davion; p 129 (b, l) Beckwith; p 129 (b, r) Davion; p 130 J. Tong; p 132 Rogers; p 133 (t, r) Pietropaolo; p 134 (t) Rogers; p 134 (b, r) Pitt; p 135 Bay Books Picture Library; p 136 Gordon Cheers; p 137 Cheers; p 138 Bay Books Picture Library; p 139 (b) J. Tong; p 140 (t, l) J. Tong; p 140 (b, r) J. Tong; p 141 J. Tong; p 142 (b, l) Bogner; p 143 (b, l) Bütschi; p 143 (t, r) Bütschi; p 144 (b, l) Beckwith; p 153 Sandved; p 155 Bogner; p 160 Pitt.

INDEX

This index is based on scientific names; common names are cross-referenced. Page numbers in *italics* refer to photographs and illustrations.

Aldrovanda vesiculosa 1, 49, 49-50, *50*
   classification of 21
   cultivation 150
   evolution of *16*
   monthly calendar 146-9
   trapping mechanisms 24-5
Amorphophallus riviera 7

Beaufortica sparsa 132
Biovularia 123
bladderwort
   alpine *see Utricularia alpina*
   sun *see Utricularia chrysantha*
   yellow *see Utricularia australis*
Brocchinia 2, 51-2
   hectioides 51
   reducta 51, *51*, 52, *52*, 143
      classification of 21
      monthly calendar 146-9
   steyermarkii 143
bugle grass *see Sarracenia oreophila*
butterwort
   alpine *see Pinguicula alpina*
   common *see Pinguicula vulgaris*
   pale
      see Pinguicula alpina, Pinguicula lusitanica
Byblis 1, 53-4
   classification of 21
   evolution of 15, *16*
   gigantea 53, 54, *54*, 133
      cultivation 29-31, 32, *32*, 150
      monthly calendar 146-9
   in a glasshouse 46-7
   liniflora 53, 54, *54*
      cultivation 32, 150
      monthly calendar 146-9

Captosis berteroniana 2, *18*, *55*, 55-6, *56*
   classification of 21
   monthly calendar 146-9
Cephalotus follicularis 1, 7, 11, *20*, *57*, 57-8, *58*, *59*, 133
   Albany, WA 131-3
   classification of 21
   cultivation *32*, 32-3, 150
   evolution of *17*
   in a glasshouse 46-7
   monthly calendar 146-9
   propagation 39
   responses to stress 15
   trapping mechanisms 23, 25-6, *26*
cobra lily *see Darlingtonia californica*
cruel plant *see Physianthus albens*

Darlingtonia californica 1, 11, 60-1, *61*
   classification of 21
   cultivation 34, 150
   evolution of *17*
   monthly calendar 146-9
   propagation 39-40, *40*
   tissue culture 44-5
   trapping mechanisms 26
dew thread *see Drosera filiformis*
Dionaea muscipula 1, 11, *18*, *62*, 62, *63*, 63, 144
   classification of 21
   containers for 29

cultivation 29-31, *33*, 33-4, 150
evolution of *16*
in a glasshouse 46-7
grafting 42
monthly calendar 146-9
North America 144
propagation 40, *40*
responses to stress 15
seeds 30
tissue culture 44-5
trapping mechanisms *23*, *24*, 24-5, *25*
Drosera 1, 2, 11, 64-81
   adelae 65, *65*
   Albany, WA 131-3
   aliciae 65, *65*
   andersoniana 66, *66*
   Anglesea, Victoria 133
   arcturi 1, 135, 137
   Arthur's Pass National Park, NZ 135
   binata 66, *66*, *134*, 135
   binata dichotoma 66
   binata 'multifida' 66
   binata 'T' form 66
   brevifolia 67, *67*
   bulbosa 133
   capensis 64, *67*, 67
   capillaris 68, *68*, 144
   cistiflora 68, *68*
   classification of 21
   containers for 29
   cultivation 29-31, 34-5, *35*, 150
   dichrosepala 69, *69*, 133
   early work on 9
   ericksonae 69, *69*
   erythrorhiza 70, *70*
   erythrorhiza ssp *colina* 70
   erythrorhiza ssp *erythrorhiza* 70
   erythrorhiza ssp *magna* 70
   erythrorhiza ssp *squamosa* 70
   evolution of 15, 16
   filiformis 19, 70, *70*
   filiformis form *filiformis* 70
   filiformis var. *tracyi* 70
   forms of *2-3*
   gigantea 71, *71*, 133
      cultivation 35
   glanduligera 71, *71*, 135
   in a glasshouse 46-7
   hamiltonii 72, *72*, 132
   helodes 30, *81*
   intermedia 72, *72*, 144
   leucoblasta 73, *73*
   linearis 73, *73*
   macrantha 133
   macrantha ssp *planchonii* 74, *74*
   menziesii 133
   menziesii ssp *menziesii* 74, *74*
   microphylla 74, *74*
   modesta 75, *75*
   monthly calendar 146-9
   myriantha 132, 133
   oreopodion viii, 48
   paleacea 20, *38*, 75, *75*, 133
   parvula 76
   peltata 1, *19*, 76, 134
   peltata ssp *auriculata* 76
   peltata ssp *peltata* 76
   platypoda 133
   platystigma 133

   prolifera 1, 64, 77, *77*
   propagation 41
   pulchella 77, *77*, 133
   pygmaea 78, *78*, 133, 135
   roraimae 143
   rotundifolia 22
   spathulata 79, 135-8, *136*
   stolonifera 133
   stolonifera ssp *compacta* 80, *80*
   stolonifera ssp *humilis* 80
   stolonifera ssp *rupicola* 80
   stolonifera ssp *stololifera* 80
   subhirtella 80, *80*, 133
   subhirtella var. *Moorei* 80
   tissue culture 44-5
   villosa 14
   whittakeri 81, *81*, 134
   whittakeri var. *praefolia* 81
Drosophyllum lusitanicum 82-3, *83*
   classification of 21
   cultivation 35, 150
   evolution of *16*
   monthly calendar 146-9

Errienellam see Drosera peltata ssp *auriculata*

fairy aprons *see Utricularia dichotoma*
flycatcher *see Sarracenia alata*
frog bonnets *see Sarracenia oreophila*

Genlisea 1, 84-5
   classification of 21
   containers for 29
   cultivation 29-31, 150
   evolution of 17
   guianensis 85, *85*
   hispidula
      trapping mechanisms 26
   monthly calendar 146-9
   repens 85, *85*
   roraime 143
   trapping mechanisms 26-7
Gronovia scandens 7

Heliamphora 1, 11, 86-9
   classification of 21
   cultivation 35-6, 150
   evolution of 15, *17*
   heterodoxa 18, *87*, 87
   heterodoxa var. *exappendiculata* 86, 87
   heterodoxa var. *glabra* 87
   heterodoxa var. *heterodoxa* 87
   minor 88
   monthly calendar 146-9
   nutans 86, *86*, 143
   propagation 41
   tatei 89
   tatei var. *macdonaldae* 89
   tatei var. *neblinae* 89
huntsman's cap *see Sarracenia purpurea*

Ibicella lutea 2
   classification of 21

Leptospermum firmum 132
lobster pot *see Sarracenia psittacina*

marsh pitchers *see Heliamphora*
monkey's rice pot *see Nepenthes albo-marginata*

Nepenthes 1, *20*, 90-107
   aerial layering *43*, 43
   alata 90, 92, *92*

albo-marginata 92, *92*
ampullaria 91, 93, *93*
ampullaria x *gracilis* 93
ampullaria x *rafflesiana* 93
x *henryana* 106
bicalcarata 93, *93*
burbidgeae 10, 94, *94, 140,* 141
burkei 10
classification of 21
courtii 10
cultivation 36, 150
dominii 10
early work on 9–12, *11, 15*
evolution of 15, *17*
formosa 107
fusca 94, *94, 95*
in a glasshouse 46–7
gracillima 95, *95*
grafting 42, *42*
gymnamphora 95, *95*
hookeriana 10, 36
khasiana 96, *96*
leptochila 96, *96*
lowii 19, 97, *97*
macfarlanei 98, *98*
masoalensis 19
maxima 13, 98, *98, 99*
  propagation *44*
mirabilis 1, 100, *100*
mixta 106
monthly calendar 146–9
Mount Kinabalu, Borneo 139–42
pilosa 100, *100*
propagation 41
rafflesiana 8, 91, 101, *101*
rajah 1, 20, 101, *101, 141*
reinwardtiana 102, *102*
responses to stress 14
x *rokko* 106
rufescens 12
sanguinea 102, *102*
sedeni x *hookeriana* 13
seeds 30
stenophylla 103, *103*
x *superba* 13
tentaculata 1, 103, *103,* 141
truncata 104, *104*
veitchii 10, *12,* 104, *104*
ventricosa 105, *105*
villosa 18, 105, *105,* 141

Physianthus albens 6
Pinguicula 108–14
  agnata 109, *109*
    cultivation 37
  alpina 109, *109*
  caerulea 144
  classification of 21
  colimensis 110, *110*
  containers for 29
  cultivation 29–31, 37, 150
  esseriana 110, *110*
  evolution of 15, *16*
  grandiflora 108
  longifolia 111, *111*
  lusitanica 111, *111*
  lutea 108, 144
  monthly calendar 146–9
  moranensis 8, 112, *112*
  primuliflora 108, 113, *113*
  pumila 112, *112*
  rotundiflora 31, 113, *113*
  vallisneriifolia 108

  vulgaris 108, 114, *114*
    cultivation 37
  zecheri 114, *114*
pink petticoats *see Utricularia multifida*
pitcher plant
  Albany *see Cephalotus follicularis*
  Californian *see Darlingtonia californica*
  hooded *see Sarracenia minor*
  marsh *see Heliamphora*
  old pale *see Sarracenia alata*
  painted *see Nepenthes burbidgeae*
  parrot *see Sarracenia psittacina*
  sun *see Heliamphora*
  sweet *see Sarracenia rubra*
  tropical *see Nepenthes mirabilis*
Polypompholyx 123

rainbow
  modest *see Drosera modesta*
  purple *see Drosera microphylla*
rainbow plant *see Byblis liniflora*
redcoat *see Utricularia menziesii*

Sarracenia 1, 11, 47, 115–20
  alata 115, *115*
  x *catesbeyi* x
  classification of 21
  containers for 29
  cultivation 29–31, *30,* 37, 151
  evolution of *17*
  flava 20, 27
    propagation *41*
  in a glasshouse *46,* 46–7
  leucophylla 28, 116, *116*
    propagation 41
    trapping mechanisms 23
  minor 37, 116, *116*
  monthly calendar 146–9
  oreophila 19, 117, *117*
  propagation 41
  psittacina 115, 117, *117*
    cultivation 37
    propagation 41
    trapping mechanisms 23
  purpurea 18, 115, 118, *118*
    cultivation 37
    propagation 41
    trapping mechanisms 23
  purpurea purpurea 118
  purpurea x *rubra* 27
  purpurea ssp *purpurea heterophylla* 118
  purpurea ssp *venosa* 118
  responses to stress 14
  rubra 20, 119
  rubra ssp *alabamensis* 119
  rubra ssp *gulfensis* 119, *120*
  rubra ssp *jonesii* 119
  rubra ssp *purpurea* 119
  rubra ssp *rubra* 119
  rubra ssp *wherryi* 119, *120*
  seeds 30
  trapping mechanisms 23, 27
  triffids based on 6
Sauromatum guttatum 6
sidesaddle plant *see Sarracenia purpurea*
sun pitchers *see Heliamphora*
sundew
  alpine *see Drosera arcturi*
  bright *see Drosera ericksonae*
  cape *see Drosera capensis*
  climbing *see Drosera macrantha* ssp
    planchonii
  common scarlet *see Drosera glanduligera*

  cone *see Drosera parvula*
  dwarf
    see Drosera brevifolia, Drosera paleacea
  forked *see Drosera binata*
  giant *see Drosera gigantea*
  lance-leaved *see Drosera adelae*
  leafy *see Drosera stolonifera* ssp *compacta*
  orange *see Drosera leucoblasta*
  pale *see Drosera peltata*
  pink *see Drosera capillaris*
  Portuguese *see Drosophyllum lusitanicum*
  pretty *see Drosera pulchella*
  pygmy *see Drosera pygmaea*
  redink *see Drosera erythrorhiza*
  rosy *see Drosera hamiltonii*
  rusty *see Drosera dichrosepala*
  scented *see Drosera whittakeri*
  spoon-leaf *see Drosera spathulata*
  sturdy *see Drosera andersoniana*
  threadleaf *see Drosera filiformis*
  trailing *see Drosera prolifera*
sunny rainbow *see Drosera subhirtella*

Triphyophyllum peltatum 2, 7–8, *13, 121,* 121–2, *122*
  classification of 21
  cultivation 29–31, 151
  evolution of 15, *16*
  monthly calendar 146–9

Utricularia 1, 11, 123–9, *133*
  alpina 124, *124*
  Arthur's Pass National Park, NZ *135*
  australia 125, *125*
  ceratophylla 126
  chrysantha 125, *125*
  classification of 21
  containers for 29
  cornuta vii, 125, *125*
  cultivation 29–31, 151
  dichotoma 126, *126*
  evolution of *16*
  fontana see Utricularia tricolor
  fusiformis see Utricularia tricolor
  gibba 123, 126
  gibba ssp *exoleta* 126
  in a glasshouse 46–7
  globularifolia see Utricularia tricolor
  gomezii see Utricularia tricolor
  hookeri 133
  humboldtii 51
  inflata 126, *126,* 144
  livida 127, *127*
  menziesii 127, *127, 133*
  monanthos 135, 137, 138
  montantha see Utricularia tricolor
  monthly calendar 146–9
  multifida 39, 123, 128, *128, 133*
  pubescens 128, *128*
  purpurea 1, 144
  quelchii 31
  rotundifolia see Utricularia tricolor
  sandersoni 129, *129*
  simplex 133
  trapping mechanisms 27
  tricolor 129, *129*
  volubilis 133
  vulgaris 18

venus fly trap *see Dionaea muscipula*
voodoo lily *see Sauromatum guttatum*

waterwheel plant *see Aldrovanda vesiculosa*
winged nepenthes *see nepenthes alata*

yellow trumpet *see Sarracenia flava*